Your Conscious Mind

Your Conscious Mind

Unravelling the greatest mystery of the human brain

NEW SCIENTIST

New Scientist

Contents

Series introduction

New Scientist's Instant Expert books shine light on the subjects that we all wish we knew more about: topics that challenge, engage enquiring minds and open up a deeper understanding of the world around us. *Instant Expert* books are definitive and accessible entry points for curious readers who want to know how things work and why. Look out for the other titles in the series:

The End of Money
How Evolution Explains Everything about Life
How Your Brain Works
Machines that Think
The Quantum World
Where the Universe Came From
Why the Universe Exists

Contributors

Editor: Caroline Williams is a UK-based science journalist and editor. She is a *New Scientist* consultant and author of *Override: My quest to go beyond brain training and take control of my mind* (Scribe, 2017).

Series editor: Alison George is *Instant Expert* editor for *New Scientist*.

Articles in this book are based on talks at the 2016 *New Scientist* masterclass on consciousness and articles previously published in *New Scientist*. They are authored by a range of experts.

Marc Bekoff is emeritus professor of ecology and evolutionary biology at the University of Colorado, Boulder, whose research focuses on animal behaviour and cognition. He wrote the section in Chapter 10 arguing that animals are conscious and should be treated as such.

Patrick Haggard is professor of cognitive neuroscience at University College London. His research focuses on the subjective experience of voluntary action and on the brain's representation of the body. He wrote the article on free will in Chapter 4.

Nicholas Humphrey is a theoretical psychologist based in Cambridge, UK, who studies the evolution of intelligence and consciousness. He is author of *Soul Dust: The Magic of Consciousness* and co-wrote 'The invention of consciousness' in Chapter 3.

J. Kevin O'Regan is former director of the Laboratoire de Psychologie de la Perception at the Université Paris Descartes who has proposed a new way to understand consciousness. He is author of *Why Red Doesn't Sound Like a Bell: Understanding the feel of consciousness* (OUP, 2011) He wrote 'Can we build "feel" into machines?' in Chapter 6.

Liz Paul is senior research fellow at the University of Bristol where she investigates the emotional and cognitive capacities of a range of animal species. She wrote about animal consciousness in Chapter 10.

Anil Seth is co-director of the Sackler Centre for Consciousness Science at the University of Sussex, UK, who researches the brain basis for consciousness. He is author of the forthcoming title, *The Presence Chamber* (Faber & Faber, 2019). His research focuses on understanding the biological basis of consciousness, which he writes about in Chapter 2 of this book.

Max Tegmark is professor of physics at the Massachusetts Institute of Technology and specializes in precision cosmology. In his book *Our Mathematical Universe* he explores the physics of consciousness, and he wrote 'Is consciousness a state of matter?' in Chapter 3 of this book.

Adam Zeman is a clinician who researches cognitive and behavioural neurology, including neurological disorders of sleep, at University of Exeter Medical School. He is author of *Consciousness: A user's guide* and writes about disorders of consciousness in Chapter 5.

Thanks also to the following journalists and editors:

Anil Ananthaswamy, Celeste Biever, Michael Brooks, Linda Geddes, Hal Hodson, Valerie Jamieson, Dan Jones, Kirstin Kidd, Graham Lawton, Tiffany O'Callaghan, Sean O'Neill, David Robson, Laura Spinney, Kayt Sukel, Helen Thomson, Prue Waller.

Introduction

Of all the mysteries of human existence these have to be the biggest: What is consciousness? Is it real or just an illusion? And either way, how does it work?

People have been pondering these kinds of questions since long before we knew that the brain was the place where thinking happens. Not until the 5th century BC, when Hippocrates noticed that people with brain injuries lost aspects of their consciousness, did anybody realize that it had anything to do with the brain.

But the questions didn't stop there. How can the squishy, tofu-like matter of the brain give us such richness of experience? How can we tell whether my experience is anything like yours? Or indeed if either of us are experiencing consciousness in the first place? What happens in the unconscious and how does it affect our notions of free will?

We do not yet have all the answers, and these questions will keep scientists and philosophers busy for a few more centuries yet.

What there are, though, are some fascinating ideas – many of them stranger than fiction. To navigate through the deep waters of philosophy and neuroscience we have brought together the ideas of the greatest minds in consciousness research and combined them with the expertise of *New Scientist* writers. We admit that the following pages do not hold all the answers to the mysteries of our minds, but they will certainly raise some fascinating new questions. They may even make you rethink everything you thought you knew about reality.

Caroline Williams, Editor

I

An introduction to the hard problem of consciousness

There are a lot of hard problems in the world, but only one of them gets to call itself 'the hard problem'. And that is the problem of consciousness — how a kilogram or so of nerve cells conjures up the seamless kaleidoscope of sensations, thoughts, memories and emotions that occupy every waking moment.

The enigma of consciousness

Ask yourself this: Do you feel conscious? The fact that you are even able to consider such a question suggests that the answer is probably 'yes'. Our own consciousness seems to be such an obvious feature of our lives that most of the time we never even stop to ponder it.

Now take a look into the eyes of the nearest human being. Are they conscious too? This time it's much more difficult to be sure. It doesn't matter whether you are gazing into the eyes of your beloved or a complete stranger; there is no way of truly knowing whether they are conscious, too. And even if they are, it's impossible to know whether their experience of consciousness is anything like yours. Start trying to ask the same sorts of questions of animals and even machines and things start to get even more complicated.

These basic features of understanding consciousness have had philosophers scratching their heads for centuries. Back in the 17th century, René Descartes set the tone for the modern debate about the problem by proclaiming that the body and conscious mind are cut from very different kinds of cloth. In Descartes' view, the body and the brain are made of matter in the same way as other physical objects such as tables and chairs, and rocks and plants. The mind, however, with our thoughts, beliefs, mental lives and memories, is immaterial – something that can neither be seen nor touched nor directly observed. This observation has set the tone for much of the debate about consciousness since.

The hard problem

In 1995 philosopher David Chalmers, at New York University, updated Descartes' point of view, dubbing it 'the hard problem'. Chalmers argued that understanding how the brain works doesn't tell you anything about consciousness because while the brain physically exists, the contents of the conscious mind cannot be observed or measured.

Understanding the brain, in Chalmers's view, is 'the easy problem'. We can tell, for example, that the brain is made up of a kilogram or so of highly connected nerve cells, some of which are specialized for certain functions. We can also tell that the currency of communication between nerve cells is both electrical and chemical. But while we can explain, for example, how our eyes inform our brains about the wavelength of light that relates to a colour, this doesn't tell you anything about what it is like to see the colour red. In this view, even understanding every detail about the brain's functioning doesn't help us understand consciousness because it tells you nothing about what it is 'like' to experience red. Or, as Thomas Nagel, also at New York University, put it in the 1970s: you could know every detail of the physical workings of a bat's brain, but still not know what it is like to be a bat (*see* box: Qualia).

Another example. Take this book in front of your eyes. Right now you are presumably having a conscious experience of seeing the paper (or screen), the words and the pictures. The way you see the page is unique to you, and no one else can know exactly what it is like for you. This is how consciousness is defined: it is your own private, personal and highly subjective experience and there is no way to explain the sense of what it is like for you to anyone else.

Qualia

In philosophical terms the 'what it's like' of our experiences are called **qualia**. These are the subjective, personal qualities of an experience: the coolness of water, the redness of red, the feeling of happiness. Proponents of the hard problem argue that no amount of understanding of the brain's physiology will properly describe qualia because there are as many versions as there are people in the world and no way of comparing them. Indeed, some have suggested that the qualia of conscious experience might be impossible to understand within our current understanding of the laws of physics.

So, if consciousness isn't a physical thing, what is it? The extreme version of this view is that consciousness is a fundamental component of the universe, one that exists alongside matter and has properties which, perhaps conveniently, cannot be explained by our present understanding of physics. If taken to the extreme, says Chalmers, this idea can lead to **panpsychism**, the view that all matter – even inanimate objects like rocks – is imbued with some degree of consciousness.

Zombies

Another challenge is that it is impossible to know whether another being is experiencing qualia at all. It is possible that everyone else is a 'zombie'. Not in the horror movie meaning of the word – zombies, of the kind found in philosophical thought experiments, are people who behave almost exactly like everybody else except for one crucial difference: they are not conscious. Stick this zombie with a pin and it will say 'ouch' and recoil. But that's just a reflex – it feels no pain. In fact, this zombie has no subjective sensory experiences, or 'qualia', at all. No one has yet found a way to tell for sure that the people around us aren't zombies.

The not-so-hard problem

At the other end of the scale, 'materialists' like philosopher Daniel Dennett, of Tufts University in Medford, Massachusetts insist that there is no such thing as the hard problem, and that ultimately we'll be able to understand consciousness – and perhaps find a way to measure qualia and spot zombies – when we understand enough about the way the brain works.

For Dennett, there is no mysterious process required for the brain's information-processing capabilities to become conscious. In fact, he calls Descartes' ideas 'one of the greatest mistakes in the history of thinking'.

Dennett argues that consciousness is a direct result of the workings of the brain. In this view, the brain is a kind of hypothesis-making machine, constantly throwing up new 'drafts' of what is going on in the world and updating them on the fly. The resulting consciousness, then, isn't some mysterious out-of-body

FIGURE 1.1 Only you can know exactly what your mind is experiencing

experience, but a by-product of the flow of information in the body and brain. In other words, a very convincing illusion.

What's more, the brain doesn't just create the illusion of consciousness but also the feeling that there is a separate, immaterial 'I' having a conscious experience. This, too, can be viewed as either a mysterious 'other' state that defies explanation or another illusion, stitched together from our life experiences and our relationships with others.

While there are no simple answers to any of these hard questions of consciousness, from a scientific point of view the materialists' theory has two advantages.

First, there is no need to explain strange interactions between material and immaterial things because in the materialist point of view what seems to be immaterial is nothing more than smoke and mirrors. And, second, it makes the hard problem disappear in favour of a drive to explain how the brain accomplishes this trickery.

Over the past two decades this has brought the problem into the realms of neuroscience. Read on to find out what this line of enquiry has taught us so far.

2

The biological basis of consciousness

Neuroscientists have made incredible progress in understanding the biological basis of consciousness, and thanks to technological advances can now even watch it in action in the brain. Here is a primer.

The raw material of consciousness

The basis of consciousness in the brain is mysterious, but at least it's an accessible mystery. As the author Mark Haddon put it recently, the raw material of consciousness is not on the other side of the universe, it didn't happen 14 billion years ago and it's not squirrelled away deep inside an atom. It's right here, inside your head.

In fact, if we put aside the philosophical question of why consciousness exists at all, we can begin to probe the brain for its tell-tale physical and electrical signatures, the so-called **neural correlates of consciousness**.

Unfortunately, the brain doesn't give up its secrets easily. The brain contains, at last count, nearly 90 billion neurons with so many connections between them that if you counted one every second it would take you three million years to complete the task. But even this doesn't do justice to the complexity of the brain. What's truly extraordinary is not the structure itself, but the patterns of connectivity that flow through it which, somehow, underlie everything that makes you, you.

How these patterns of connectivity add up to consciousness is a huge question. So where should we begin in our attempts to understand how it all works? One approach is to break the problem down into manageable chunks, and to explore the biological basis of the different aspects of consciousness, one at a time.

We can, for example, differentiate between the *level* of consciousness (the difference between being vividly awake and aware or under general anaesthesia), the *content* of the experience (what it is that we are sensing and responding to) and the conscious sense of *self* (the mysterious but at the same time

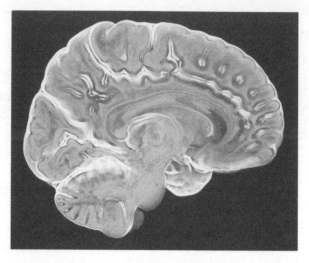

FIGURE 2.1 In theory, we could look inside the brain and measure consciousness in action

completely familiar feeling that everything is being experienced by a unified 'me').

Level of consciousness

What in the brain determines whether we are conscious or not? At the simplest level the brain does seem to have at least one on/off switch. The **intralaminar thalamic nuclei** are part of the **thalamus**, which sits in the very centre of the brain, at the top of the brainstem. If this part of the brain is damaged, it will shut off consciousness entirely. The **claustrum**, a thin sheet of tissue deep inside the brain, also seems to have an important role in whether we are conscious, or awake but unconscious (*see* box).

A sweet spot for consciousness?

One moment you're conscious, the next you're not. Can there really be such a thing as an on/off switch for consciousness in the brain? It seems so. In 2014 researchers were able to switch a woman's consciousness on and off by stimulating one small region of her brain.

The woman, who was undergoing exploratory surgery to localize the source of her epileptic seizures, had an electrode inserted next to a thin sheet of brain tissue called the claustrum, hidden deep within the brain. This was a region that had never been stimulated before.

When the team zapped the area with high-frequency electrical impulses, the woman lost consciousness. She stopped reading and stared blankly into space, she didn't respond to auditory or visual commands and her breathing slowed. As soon as the stimulation stopped, she immediately regained consciousness with no memory of the event.

Although still only tested in one person, the discovery provides evidence for the idea that the claustrum is important for making consciousness out of the maelstrom of information in the brain. Christof Koch of the Allen Institute for Brain Science in Seattle, a proponent of this idea, believes that the claustrum acts as a kind of conductor of consciousness, which integrates information across distinct regions of the brain and binds together information arriving at different times. In 2017 this theory gained further support with the discovery of three long neurons that emanate from the claustrum and encircle the mouse brain, taking in many important areas along the way.

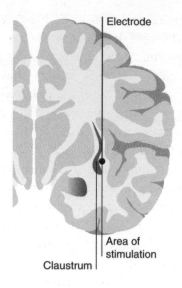

Electrode

Area of
stimulation

Claustrum

FIGURE 2.2 Located deep in the brain, the claustrum could bind our
perceptions into a cohesive whole

Everyone agrees, though, that there is more to consciousness
than a simple distinction between on and off. We know, for
example, that a person can be asleep yet still be having a similar
experience in a dream to normal, waking consciousness. On
the other hand, someone in a persistent vegetative state may be
physiologically awake, yet lacking any signs of consciousness.

The emerging picture is that while there are several important brain areas and cell types that are involved in consciousness,
the overall experience depends on the ways that activity is coordinated across the brain and the timescale on which it happens.

So how do we quantify the level of consciousness? One
very promising method comes from Marcello Massimini
at the University of Milan, Italy. He and his colleagues have
developed a method that involves stimulating brains with an

electromagnetic pulse (using so-called 'transcranial magnetic stimulation' or TMS) and then measuring how waves of activity propagate across the brain. This is measured using EEG – a measure of the brain's electrical activity that is recorded via electrodes on the surface of the skull. The pulse acts like striking a bell, and neurons across the entire brain continue to 'ring' in a specific wave pattern, depending on how active the connections between individual brain cells are.

By analysing the complexity of these wave patterns of the brain's response, Massimini and his team came up with a number between zero and one, which they called the Perturbational Complexity Index (PCI). People in a vegetative state who are unresponsive and probably unconscious have a PCI closer to zero. According to one study there appears to be a cut-off at about 0.3, which seems to distinguish conscious states from unconscious states.

Subsequent studies have used EEG measurements alone – without the pulse of electromagnetic stimulation – to see if complexity measures can still be used to determine the level of consciousness. Put very simply, these measures quantify how 'diverse' or 'unpredictable' the brain signals are. And, indeed, it turns out that these measures of spontaneous complexity also reduce from that seen in wakeful rest, through mild sedation, to full general anaesthesia. Similarly, studies in people who have electrodes implanted in their brains to help localize their epileptic seizures have shown a general decrease in complexity as people fall asleep. Interestingly in REM sleep, when people are dreaming, the complexity of their brain dynamics is much the same as it is during normal conscious wakefulness – which tells us that these measures of complexity are reflecting levels of consciousness specifically, not simply physiological changes in states of wakefulness.

As for 'higher' states of consciousness, some recent research has used a method called MEG (magnetoencephalography, which measures the tiny magnetic fields generated during brain activity) to study the brain dynamics of people under psychedelic drugs such as LSD, psilocybin and ketamine. Compared to a baseline state, these drugs seem to do the opposite of anaesthesia or falling asleep. They seem to *increase* the level of brain complexity – the first time this has been observed. Could this be a sign of having reached 'higher ground'?, an elevated level of consciousness? It's too early to say for sure, but it is an intriguing avenue for future research.

These ways of measuring consciousness level are related to an increasingly popular theory of consciousness called **integrated information theory,** or IIT for short, developed by the neuroscientist Giulio Tononi at the University of Wisconsin (*see* box: Integration breeds awareness?). However, existing measures like those mentioned above provide only crude approximations to the theory. The full version of integrated information remains virtually impossible to measure for any real system.

Integration breeds awareness?

We don't experience colours, shapes and sounds separately, but as a fully integrated whole. Giulio Tononi, a neuroscientist in Madison, Wisconsin, has put forward a theory that describes this process. He says that for a system to be conscious, it must integrate information in such a way that the whole generates more information than the sum of its parts. In conscious minds, integrated information cannot be reduced into smaller components.

When you perceive a red triangle, the brain cannot register the object as a colourless triangle plus a shapeless patch of red.

Tononi calls the measure of how a system does this, **phi**. According to his theory, the ability to integrate information is a key property of consciousness. A digital camera has a prodigious memory capacity but its millions of pixels never 'see' a photo. Your mind can because your brain actively integrates information to make sense of the data.

One way of calculating phi involves dividing a system into two and calculating how the parts predict their future state, compared to the whole system. One cut would be the 'cruellest', creating two parts that are the most independent. If the parts of the cruellest cut are completely independent, so that the 'whole' is no greater than their sum, then phi is zero, and the system is not conscious. The greater their dependency, the greater the value of phi and the greater the degree of consciousness of the system.

Tononi's approach can explain some curious aspects of consciousness. Why do we lose consciousness when we go to sleep? He would say that this is a time when information from the brain's specialized circuits is no longer integrated. Why are brain seizures associated with a loss of consciousness? It might be because seizures overload the circuits, blocking complex informational exchange.

FIGURE 2.3 Consciousness might arise from the integration of
information in the brain

In terms of which brain regions are involved in maintain-
ing conscious level, attention has recently been drawn to a
posterior 'hot zone' in the cortex – centred on the **parietal**
and **occipital cortex**. Activity in this area seems to distinguish very reliably between conscious and unconscious states.
A study by Francesca Siclari, of the University of Wisconsin–
Madison, and colleagues provides probably the best evidence
for this to date. Instead of simply comparing waking with
sleeping – a comparison which involves many changes in
the brain and body besides the loss of consciousness – they
looked only at the brain during sleep. By waking people up
many times during each night and asking them whether they
had been dreaming of anything, they were able to compare
brain activity when people had been dreaming, with that
when they had not been having any conscious experiences
at all. This way, the overall state of the brain and the body
was the same, so that any differences they found could be
tied more closely to consciousness itself. In this comparison,

the posterior 'hot zone' appears very prominently as strongly associated with conscious experience. So prominently, in fact, that these researchers were able to accurately predict whether a person would report dreaming *before* waking them up, based only on the activity in this area.

Are babies conscious?

In adults, conscious awareness of having seen, felt or heard something is linked to a two-stage pattern of brain activity. Immediately after a visual stimulus is presented, for example, areas of the visual cortex fire. About 200 to 300 milliseconds later other areas light up, including the prefrontal cortex which deals with higher-level cognition. Some researchers think that conscious awareness kicks in only after the second stage of neural activity reaches a specific threshold.

This is easy enough to study in adults because they are able to report when they are aware, but it has been impossible to ask the same questions of babies to find out if, and how, they become conscious of something in the environment.

Sid Kouider, at the École Normale Supérieure in Paris, and his colleagues tackled this question by putting EEG caps on groups of babies aged 5, 12 and 15 months, and recording brain activity while they were shown a series of rapidly changing images. As with the adults, all of the babies in the group responded to the face with the expected two-stage pattern. But in the second stage – the activity linked to conscious awareness – the response was much slower.

The slowest and least distinct reaction was in the five-month-old babies, where there was a delay of more than one second before the second pattern appeared. At 12 months, the second stage of activity arrived 800 to 900 milliseconds after the image was displayed. The 15-month-old group showed a very similar response.

It seems that babies may have the same mechanism for consciously registering things in the world around them. It just takes a little longer for the penny to drop.

The content of consciousness

When you are conscious you're conscious of something. But while it certainly seems as if what we see, hear and feel is very real, there is good evidence to suggest that what we perceive is a kind of 'controlled hallucination' – a 'best guess' by the brain about what's causing its sensory inputs.

Consider this: the brain is locked inside a bony skull. It has no direct access to the things in the world. It has no direct access even to its body. All the brain receives are electrical signals that arrive at the brain from the different sensory organs – the eyes and the ears and so on. These signals are noisy and ambiguous, yet somehow the brain has to work out what it all means.

Back in the 19th century, the German physiologist Hermann von Helmholtz proposed that the way the brain does this is by acting as a kind of prediction machine. It combines sensory information coming from the world, with its prior assumptions (or expectations) about the way the world is. This leads to a 'best guess' about what caused the sensory signals – and this is what we consciously perceive.

FIGURE 2.4 Adelson's checkerboard illusion

This concept isn't easy to swallow at first, but the process is easy enough to demonstrate using a simple visual illusion: Adelson's Checkerboard (*see* diagram).

On first inspection, the squares marked A and B, look to be different shades of grey. But they are, in fact, the same. What's happening here is that the brain is using its prior knowledge that a shadow cast on a surface will make that surface appear darker. Combined with the fact that we expect B to be the same colour as all of the other squares in that diagonal line, the brain generates a prediction that square B is probably light grey in the 'real world'. As a result the brain perceives it as light grey, but slightly darkened by a shadow. This effect is so strong that even when you have been shown the illusion, it doesn't change the perception.

Thinking of things this way dramatically changes how we see the brain. Rather than the senses faithfully recording what is out there and informing the brain, in this view it is the connections flowing from inside the brain back out towards the sensory surfaces that are doing the perceptual heavy lifting. In other words, the content of our conscious perceptions is largely a construction of the brain: a 'controlled hallucination' in which our perceptual predictions are continually reined in by sensory signals impinging from the outside world.

As for the physical basis of this best-guessery in the brain, there is increasing evidence that perceptual predictions and top-down signalling have important influences on conscious perception. Early on, back in 2001, neuroscientists Vincent Walsh and Alvaro Pascual Leone asked people to look at clouds of moving dots, while they interrupted brain activity using **transcranial magnetic stimulation** (tMS). When they interrupted the top-down (or inside-out) signals, they found that people were no longer consciously seeing the dots move. This suggested that the brain needs its internal predictions to make sense of what is going on in the external world.

More recently, Lars Muckli and his colleagues at Glasgow University showed that you could use **functional MRI** (which measures metabolic activity or blood flow in the brain) to 'decode' what kind of visual scene people were looking at. Critically, they could decode this information even from part of the visual cortex that was not receiving any sensory input – meaning that it had to be based on 'top-down' predictions coming from other parts of the brain.

Another recent study has tied perceptual predictions to the so-called **'alpha rhythm'** in the brain. This is a prominent oscillation in brain activity (a 'brain wave'), which happens at about ten hertz – or ten cycles per second – and is especially obvious across the back of the brain, around the visual cortex. This research, carried out at the Sackler Centre at the University of Sussex, found that perceptual predictions had a bigger effect on conscious perception at a particular 'phase' of the alpha cycle: for example, whenever the wave is at its 'peak' a perceptual prediction will have a bigger effect on what a person consciously sees, than when the wave is at its trough.

What does being conscious of something allow us to do? It may seem obvious, but when we're conscious of something, we are able to behave very flexibly. If I see a coffee cup, I can ignore it, pick it up – or toss it across the room, whatever I feel like. An influential line of research – **global workspace theory** (*see* box The global workspace model of consciousness) – suggests that this flexibility happens because whatever is in our consciousness at any time is 'broadcast' widely among different brain regions, allowing the person to respond in all sorts of different ways. In fact, many proponents of global workspace theory argue that this process of 'broadcasting' is in fact what consciousness *is*.

To test the association between conscious perception and global broadcast, a useful approach is to compare the brain activity evoked by a visible stimulus (like a letter 'A' presented very clearly on a screen), with the brain activity evoked by the same stimulus when it's not consciously visible. There are many ways to do this, for example by showing the stimulus very briefly and immediately following it with a meaningless pattern – a technique known as **masking**. Many experiments following this method found that a large part of the cortex – the so-called **fronto-parietal network** – lights up when people report a conscious perception, compared to when they don't. This seems to strongly support global workspace theory, which associates the fronto-parietal network with the workspace. However, some recent findings, including those about the posterior cortical 'hot spot' mentioned earlier, are beginning to challenge this view.

The global workspace model of consciousness

The content of our conscious experience is constantly changing. The 'global workspace model', first proposed in 1988 by Bernard Baars of The Neuroscience Institute in San Diego, California, attempts to describe the mechanics of how those shifts happen.

Baars suggests that non-conscious experiences are constantly being processed locally within separate regions of the brain and that the brain also keeps track of what is going on in the body and memory. Different aspects of our ever-changing experience only become conscious when this information is broadcast to a network of neural regions, the 'global workspace' distributed across many different regions of the brain. This then reverberates in a flash of coordinated activity and we consciously register whatever it is that we are experiencing.

Support for this idea comes from so-called binocular rivalry experiments, which provide good evidence that the brain does indeed actively select which information to send to our consciousness. Normally, both our eyes see the same scene, so the brain can easily combine the two monocular inputs into a coherent picture. But present the left eye with an image that's dramatically different from what the right eye is seeing, and experiments have revealed that the brain resolves this conflict by allowing you to see only one image at any one time. In other words, you are only conscious of either the left-eye image or the right-eye image, but never both simultaneously.

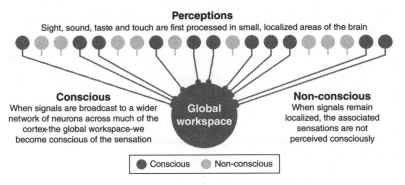

FIGURE 2.5 Broadcasting consciousness: The global workspace suggests that consciousness arises from highly coordinated, widespread activity in the brain.

Exactly what happens in the brain when we become conscious of something is still being figured out and there is much still to learn. But it is looking increasingly likely that our experience of 'reality' is actually a kind of controlled hallucination generated by the brain and updated on the fly.

Hold that thought

When it comes to consciousness, the brain may be doing just that. Recent research suggests that conscious perception requires brain activity to hold steady for hundreds of milliseconds. This signature in the pattern of brainwaves can be used to distinguish between levels of impaired consciousness in people with brain injury.

Neuroscientists think that consciousness requires neurons to fire in such a way that they produce a stable pattern of brain activity. The exact pattern will depend on what the sensory information is, but once information has been processed, the idea is that the brain should hold a pattern steady for a short period of time – almost as if it needs a moment to read out the information.

In 2009, Aaron Schurger of the Swiss Federal Institute of Technology in Lausanne tested this theory by scanning 12 people's brains with fMRI machines. The volunteers were shown two images simultaneously, one for each eye. One eye saw a red-on-green line drawing and the other eye saw green-on-red. This confusion caused the volunteers to sometimes consciously perceive the drawing and sometimes not.

When people reported seeing the drawing, the scans, on average, showed their brain activity was stable. When they said they didn't see anything, it was more variable. Schurger and his colleagues repeated the experiment – using electroencephalography and magnetoencephalography, which measure the electrical and magnetic fields generated by brain activity. These techniques provide greater temporal resolution than fMRI, allowing the team to see how the pattern of activity changes over milliseconds within a single brain.

Based on their earlier work, the team expected the volunteers' brain activity to stabilize and stay that way for hundreds of milliseconds when they reported having seen the drawing, but become highly variable otherwise.

The team then tested their technique in 116 people with disorders of consciousness. The patients, who were either minimally conscious, in a vegetative state or had just recovered from coma, were played a tone while their brain activity was recorded. The more conscious the patient, the greater the stability of their brain activity.

The work supports and augments the global neuronal workspace theory of consciousness.

The consciousness connection

An important clue in the hunt for the origin of consciousness was unearthed almost a century ago.

When Constantin von Economo peered down the lens of his microscope in 1926, he saw a handful of brain cells that were long, spindly and much larger than those around them. They looked so out of place that at first he thought they were a sign of some kind of disease. But the more brains he looked at, the more of these peculiar cells he found – and always in the same two small areas that evolved to process smells and flavours.

Von Economo briefly pondered what these 'rod and cork-screw cells', as he called them, might be doing, but without the technology to delve much deeper he soon moved on to more promising lines of enquiry.

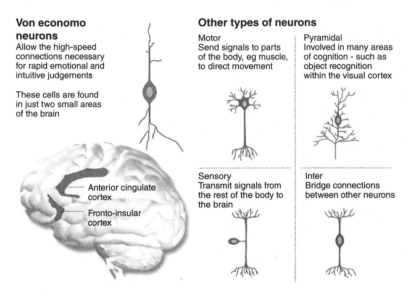

Von economo neurons
Allow the high-speed connections necessary for rapid emotional and intuitive judgements

These cells are found in just two small areas of the brain

Anterior cingulate cortex

Fronto-insular cortex

Other types of neurons

Motor
Send signals to parts of the body, eg muscle, to direct movement

Pyramidal
Involved in many areas of cognition - such as object recognition within the visual cortex

Sensory
Transmit signals from the rest of the body to the brain

Inter
Bridge connections between other neurons

FIGURE 2.6 Judgement cells. Von Economo neurons may play an important role in our sense of self

Almost 80 years later, Esther Nimchinsky and Patrick Hof, then both at Mount Sinai University in New York, also stumbled across clusters of these strange-looking neurons, and this time they had the technology to study them in greater detail. Now, after more than a decade of functional imaging and post-mortem studies, there is mounting evidence that they may have something to do with the rich inner life that we call consciousness.

These giant brain cells, now known as **von Economo neurons** (VENs), are at least 50 per cent, and sometimes up to 200 per cent, larger than typical human neurons. And while most neurons have a pyramid-shaped body with a finely branched tree of connections called **dendrites** at each end of the cell, VENs have a longer, spindly cell body with a single projection at each end with very few branches (*see* diagram). They are also rare, making up just 1 per cent of the neurons in the two small areas of the human brain: the **anterior cingulate cortex (ACC)** and the **fronto-insular (FI) cortex**.

Their location in those regions suggests that VENs may be a central part of our mental machinery, since the ACC and FI are heavily involved in many of the more advanced aspects of our inner lives. Both areas kick into action when we see socially relevant cues, be it a frowning face, a grimace of pain or simply the voice of someone we love. When a mother hears a baby crying, both regions respond strongly. They also light up when we experience emotions such as love, lust, anger and grief.

The two brain areas also seem to play a key role in the 'salience' network, which keeps a subconscious tally of what is going on around us and directs our attention to the most pressing events, as well as monitoring sensations from the body to detect any changes. What's more, both regions are active when a person recognizes their reflection in the mirror, suggesting that these parts of the brain underlie our sense of self – a key component of consciousness.

This may amount to a continually updated sense of 'how I feel now': the ACC and FI take inputs from the body and tie them together with social cues, thoughts and emotions to quickly and efficiently alter our behaviour.

The sensorimotor theory of consciousness

One theory as to the biological basis of consciousness concentrates not on the brain but on the way that we interact, bodily, with the environment. According to Kevin O'Regan of Paris Descartes University, our experience of qualia is not something that is generated by some kind of special brain mechanism, but rather simply consists in all the things that we do when we interact with the world via our senses.

When we feel a rough surface, the precise nature of the sensation is made up of all the things that happen when our fingers run over the surface: for example the vibrations in our fingers change in specific ways when we move our fingers quickly or slowly, or when we press hard or lightly as we move them. The feel of roughness is precisely constituted by all those possible patterns of interaction.

Looking at it, we can imagine the roughness, but this doesn't have the same quality of realness because we are not actually interacting physically with the surface. It is not enough to simply interact with the world, however, O'Regan argues that we also need to pay attention to a stimulus and process that information. Otherwise, however red the red and however rough the rough surface, we will never be conscious of having experienced it. See Chapter 6 for more detail on this.

The Self

It's there when we wake up and slips away when we fall asleep, perhaps to reappear in our dreams. It's that feeling we have of being anchored in a body we own and control and perceive the world from within. It's the feeling of personal identity that stretches across time, from our first memories, via the here and now, to some imagined future. The conscious self is the specific experience of being you, and the experience of being a conscious being is not one thing, but many.

We can think of the self as having three strands: the physical self (which arises from our sense of embodiment); the psychological self (which comprises our subjective point-of-view, our autobiographical memories and the ability to differentiate between self and others); and a higher-level sense of agency, which attributes the actions of the physical self to the psychological self (*see* Chapter 4 for more on this).

In looking for an explanation of how this adds up to a unified self, modern neuroscience is leaning towards a similar kind of explanation as for many other features of consciousness: that it is an illusion that emerges from other, more general purpose processes. In other words, our experience of being an integrated self is also the brain's best guess at the causes of signals that arise from our bodies, the environment and our social world.

Evidence for this comes from various quarters. In the famous rubber hand illusion (*see also* Chapter 11), for example, the experimenter uses a paintbrush to stroke a volunteer's hand (which is hidden from view) and an adjacent, visible rubber hand. The stroking is done simultaneously at the same speed and place on both the real and rubber hand. Within minutes, most people report feeling the touch of the brushstrokes on the rubber hand as if it belonged to them. This suggests that our sense of me is flexible enough to add something totally foreign to the concept of 'me'. The self is constantly being updated based on what our brain's best guess is about the source of incoming signals.

The sense of physical self

An updated version of this experiment has probed the effect further. Using a virtual reality version of the rubber hand illusion, one study showed that a volunteer was more likely to feel ownership over the virtual hand when it was flashing in time with their heartbeat than when it was flashing out of sync.

This shows that the sense of the physical self depends not just on signals from the outside (the feeling of being stroked), but also on internal measurements of 'me-ness', such as the rhythm of our heartbeat.

As for the question of where the self resides, that is tricky to answer. There are some tantalizing hints that particular kinds of neurons might be well suited to the rapid-fire integration that underlies conscious selfhood (see The consciousness connection above), but these ideas remain speculative and in general it looks like a thankless task to associate the 'self' with any single part of the brain.

What's more, the self is also constantly changing. Each time we recall an episode from our past, we remember the details slightly differently, and in doing so alter our personal history: a key aspect of the experience of being a self. The self you are today may feel solid, but it is built on shifting sands.

FIGURE 2.7 The brain regions that are active when a person recognizes their reflection in the mirror could underlie our sense of self

Altogether, many mysteries remain about how the brain and body generate the 'inner universe' of consciousness that we each enjoy.

The biological basis of subjective experience remains one of the greatest mysteries left for the next generation of scientists. This knowledge would not only help us understand what it is to be human, it would also throw new light on the nature of psychological disorders and how to fix them. And, perhaps most immediately, would offer a lifeline to those suffering from severe disorders of consciousness.

There are tens of thousands of people in a coma, vegetative state or minimally conscious state in the UK alone. Recent developments in the understanding of the brain's basis of consciousness are getting us closer to not only understanding when a patient may have even a fleeting conscious experience, but are also opening up the possibility of communication with them. And if there is one thing that makes human life worth living it is the ability to communicate your conscious experience and make yourself understood.

Timeline: Shining a light on consciousness

1641
French philosopher René Descartes distinguishes between the material self (the body) and the immaterial self (the mind).

1690
Philosopher John Locke defines consciousness as 'the perception of what passes in a man's mind', setting the tone for later research.

1968
Discovery of the four stages of sleep as we move through different levels of unconsciousness.

1960s
Roger Sperry does the first studies of 'split brain' patients, and describes strange disorders of self-awareness and perception.

1970
Gordon Gallup Jr develops the mirror self-recognition test.

1974
Thomas Nagel publishes his essay 'What is it like to be a bat?', raising the problem of subjectivity in understanding conscious experience.

2011
Adrian Owen uses EEG to enable patients thought to be in a vegetative state to respond.

1998
Rubber hand illusion first showed that our sense of self is more flexible than we previously thought.

2014
Christof Koch suggests that consciousness is a fundamental property of networks.

2014
Computer AI passes Turing Test.

1838
Charles Darwin watches an orangutan look in a mirror and ponders if she has self-awareness. Later inspires the 'mirror test' of self-consciousness.

1890
William James, philosopher, publishes *The Principles of Psychology.*

1950
Alan Turing invents the Turing Test as a standard for machine consciousness.

1924
Invention of the electroencephalogram (EEG), which measures electrical brain activity in real time, allowing a window onto the mind.

1915
Sigmund Freud declares the subconscious to be the source of human behaviour.

1977
First magnetic resonance imaging (MRI) scan of a human. The method later revolutionized the study of living brains.

1985
Invention of transcranial magnetic stimulation, which can temporarily 'knock out' certain brain areas to probe their function.

1995
David Chalmers identifies the 'hard problem' of understanding consciousness.

1991
Daniel Dennett publishes 'Consciousness Explained', setting out his materialist theory of consciousness.

3
What does it all mean?

While some features of consciousness can be understood by looking at the brain, the broader philosophical issues call for a different approach. Looking to more theoretical fields such as quantum physics and philosophy, we are finding other ways to frame the big questions.

Why the feeling of consciousness evolved

In the ongoing debate between those who think consciousness can be understood and those who think it can't, there is one thing that we can all agree on: consciousness is wonderful. We know so well the heat and redness of a fire, the sour tang of a lemon, the caress of a lover's hand. These conscious sensations lie at the core of our being, and without them we'd be duller creatures living in a poorer world. Given the choice, not one of us would prefer to be a 'zombie'.

We can also agree that consciousness is, at present, unexplained. The problem is not that we do not understand consciousness at all. Some aspects of it are relatively easy to account for in scientific terms, as we have seen in Chapter 2.

Yet all of these brain-based approaches to understanding the nature of our conscious experiences share one problem. They leave out the very thing that most of us find so baffling and indeed fascinating: the eerie feeling of 'what it's like' to be conscious, the 'qualia'. They also fail to explain why we have this qualitative dimension to experience in the first place; what value might it have for biological survival?

What is the point of qualia?

There are those who argue that qualia aren't 'for' anything – that they don't have a specific job to do in the brain. Certainly, not all mental states have this special quality attached. There is no special *feel* associated with having the thought, say, that today is Thursday. It's not *like* anything to believe it's going to rain, or to remember where you put your hat. But if qualia aren't a necessary feature of higher cognitive thoughts, why have them at all?

One clue is that the 'what it's like' quality of consciousness kicks in only at a more animal level, attached mainly, perhaps exclusively, to our experience of bodily sensations. The pain of a bee sting, the salty taste of an anchovy, the blue look of the sky – they simply couldn't be the states they are without this mysterious extra dimension.

The inexplicable nature of these experiences has led generations of scientists to side-step the question altogether. As cognitive scientist and philosopher Jerry Fodor said back in 1998: 'This is, surely, among the ultimate metaphysical mysteries; don't bet on anybody ever solving it.' Yet in more recent years it does seem that a consensus is emerging, at least about how to set the boundaries of the problem. Most theorists now accept that there are only two options that can be taken seriously. We can be *realists* about qualia – that they are real, but we don't have the physics to describe them yet (*see* box: Is consciousness a fourth state of matter?, below), or else we have to be *illusionists* – that, while they feel real, this is a trick of the mind, an illusion.

Whether realist or illusionist, qualia need to be explained. There are, after all, times when our conscious experience contains little else. A science of consciousness that leaves qualia out is not just ignoring the elephant in the room, it is ignoring the elephant that *is* the room.

Even without new laws of physics, there might be a way to characterize that elephant, by taking a fresh look about how sensations might have evolved.

Evolving inner world

First, imagine a very ancient and primitive creature floating in the sea, responding to stimuli with reflex wriggles of acceptance or rejection. These responses have been honed by evolution and are meaningful in that they contain information about what kind of stimulus is reaching the body, what part of the body is affected, and what impact it has on biological well-being. At first, though, there is no 'self' to ponder this information further.

Before long there arises in the brain a special module – a kind of proto self – whose job is to extract meaning from the motor commands that are driving the wriggling response. This is just the beginnings of sensation but at this stage there is nothing fancy or magical about the interpretation and there are no special feelings attached.

As the descendants of the original creatures evolve to be more sophisticated, these overt responses may have proved inconvenient – you might not always want to wriggle if it will give your position away to a predator. So the creature faces a problem: how to lose the bodily behaviour but keep the information about the meaning of the stimulus?

The solution was for the responses to become internalized such that the motor signals no longer reached the actual body surface, but were diverted back towards the body-map where the sense organs first project to the brain. This changed the response from an actual form of bodily expression to being a virtual one – yet still a response that the animal could use as information.

This change had a remarkable knock-on effect, setting up a feedback loop between motor and sensory regions of the brain that is capable of going round and round, catching its own tail. This means that the activity can be drawn out in time, so as to create the 'thick moment' of sensory experience. The activity stabilizes, so as to create what mathematicians call, an 'attractor' state, in which activity continues to

FIGURE 3.1 What gives rise to the inner experience of smelling a rose?

reverberate around the feedback loop after the stimulus that caused it has ceased.

Could it be that these neural echoes are what we experience when we look at the blue sky or smell a rose? Are qualia experiences that began in the senses but then took on a life of their own, becoming an unreal state that seems very real to the person experiencing them?

This theory has recently gained support. In experiments, the conscious experience of seeing seems to rely on the monitoring of activity in a loop running between the **primary sensory cortex** and areas further forward in the brain. When the **visual cortex** is stimulated a person gets a visual sensation. But if you block activity in the feedback loop on the return path from more frontal areas of the brain, people don't consciously see anything at all.

So this provides an answer to the question of *what* evolved. Can we take this one step further, and answer *why* it evolved? What was the evolutionary purpose for the vividness of these illusions?

Some will argue that qualia contribute nothing to cognition, that they have no impact at the level of biological survival, that anything our conscious minds can do they could do just as well if they weren't conscious. This would imply of course that qualia can't have evolved by Darwinian selection. Yet perhaps that is missing the point. Perhaps, rather than giving us a cognitive advantage in terms of being more intelligent or productive on the outside, the role of qualia is to make our sensory experiences bigger on the inside? In other words, they don't exist to simply keep us alive, but to make life worth living? Could it be that nature, when she invented qualia, did so that we could amaze ourselves and would strive to stay alive to be amazed even more?

As for how these experiences come to feel richer and more 'real' than any others, it could be that they appear in the mind's eye as a kind of hologram: a two-dimensional image that, nonetheless, gives the impression of a three-dimensional object. This idea takes us into the realms of **string theory**, where the holographic principle describes what happens to information that is apparently lost in black holes. It states that a two-dimensional surface can contain all the information to construct a three-dimensional world. Similarly, perhaps the four-dimensional world of conscious qualia could be an illusion generated from the surface of a three-dimensional brain?

Why only one self?

However wonderful our conscious sense of the world may be, none of this explains another mystery of consciousness. Why does our experience happen to just one self and not more than one? At any point in time we might be experiencing a variety of mental states, from pain in your back to a memory of your mother's face, but there is never any doubt that the same 'I' is experiencing them all.

We might think it obvious that it has to be so. But it is quite plausible to imagine that your brain could house several independent yous, each representing a different segment of the mind. Indeed, this fragmented state may have been the way we all started out at birth. During the first few months of life a baby's experience might consist of one 'me' that wiggles a toe, another 'me' that sees, another that feels hunger, and so on. At the beginning there is little cross-talk between them, but as the baby begins to interact with the outside world, these separate experiences begin to merge into a single self.

This binding of the selves would not necessarily need to be genetically pre-programmed. Instead, it might emerge automatically out of the interaction of various systems in the body. Something like this has long been known to happen in

FIGURE 3.2 Why does our inner experience happen to just one self, not more?

inanimate objects. In the 17th century Christiaan Huygens, the inventor of the pendulum clock, observed that when two or more of his clocks were hung from the same beam, their pendulums would spontaneously begin to beat in synchrony. In a more recent demonstration, a set of five metronomes are placed on a floating table, and they too soon begin beating as one. It happens because each individual metronome, interacting via the table, feels the pull of the others. Perhaps the separate parts of a newborn mind, interacting with a single body, also somehow feel the pull of the others.

This 'one-ness' brings new advantages for any creature fortunate enough to experience it. It creates a mind-wide forum that the cognitive scientist and artificial intelligence pioneer Marvin Minsky called 'The Society of Mind'. Once information from different modules has been brought to the same table it creates a useful mental space for planning and decision-making, overseen by an onboard autopilot that we call 'I'.

We can think of this 'autopilot' as working in a similar way to the automated versions that help keep aeroplanes in the sky and that are being developed to control driverless cars. The cockpit of a plane features a variety of independent instruments that monitor the internal and external states: speed, altitude, fuel reserves, global position, intended course, and so on. The pilot's job is to integrate all this information, so as to decide what to do to achieve certain goals: to observe, then think, then act. The pilot itself doesn't need to be conscious: modern autopilot technology is more than capable of integrating the different sources of information to get from A to B safely, while recording salient information in the black box recorder.

Engineers are working on driverless cars that can predict the movements of other cars on the road. If such systems can be engineered into a machine, then it should not be surprising that the brain can do it too. Some of the leading theories of consciousness explain how the brain achieves it, notably the global neuronal workspace theory and Giulio Tononi's model of integrated information. Christof Koch and Francis Crick have identified a brain structure, the **claustrum**, as master of ceremonies (*see* Chapter 2).

An important side-benefit of having a single overseeing 'I' is that it makes it possible to make sense of, and reflect on, your own experience. This supports another important function of consciousness, which is to appreciate how your own mind works. Observing how beliefs and desires generate wishes that lead to actions, the mind seems to have a clear structure. You begin to get insight into why you act the way you do. This means that you can not only explain yourself to others but – equally important – it provides a model for explaining other people to yourself (*see* box: It's not about you). In this way, consciousness lays the ground for what psychologists call **theory of mind**: the knowledge and understanding that someone else has their own point of view.

Put these two things together: the united self and the internally generated qualia of experience, and what do you get? A conscious self that is aware of how amazing its inner world really is. It might even create the world around it (*see* Does consciousness create reality?). It all begins with qualia, take them away and this remarkable source of connection that underlies human society and the whole thing would begin to fall apart.

It's not about you

Although our conscious experience feels intensely personal to us, psychologists Peter Halligan of the University of Cardiff and David Oakley of University College, London, argue that it exists to protect the wider social group, not just the individual.

The emergence of consciousness, they believe, came alongside other developments in brain processing that, together, made it advantageous to communicate our internal thoughts to others.

In order for this to happen, it was necessary to generate a personalized construction of self and attribute to it the essential cognitive abilities of awareness and agency, as well as the creation of inner perceptions of the world. In this view, it is our capacity to tell others of the contents of our consciousness that confers the evolutionary advantage – not the experience of consciousness itself.

Why should this ability be advantageous? Well, it lets you share with other people, via unconsciously driven systems, selected contents from your consciousness, including beliefs, prejudices, feelings and decisions. This, in turn, enables the development of adaptive strategies such as predicting the behaviour of others, which could be beneficial to species survival.

This sharing of unconsciously generated, consciously experienced self-narratives also allows for the possibility that the mental content of individuals can be changed by outside influences such as education and other forms of socializing. This is important for disseminating ideas regarding norms and values. In fact, Halligan and Oakley

argue that none of the social systems that human societies depend on would be possible without our compelling
sense of self awareness.

This new understanding of consciousness as serving
the needs of the social group rather than the individual
allows us to move from seeing ourselves not as individuals but as 'dividuals' whose interests and personhood are
shared with others. As the German philosopher Friedrich Nietzsche noted, 'consciousness is really only a net
of communication between human beings… consciousness does not really belong to man's individual existence
but rather to his social or herd nature'. Consciousness,
therefore, provides a powerful evolutionary advantage
by allowing shared communication, and extending each
individual's understanding of the world.

Does consciousness create reality?

Our conscious reality certainly feels real to us, but could it
actually be the thing that creates reality in the first place?
Some physicists think so.

In quantum physics a particle, such as an electron or pho-
ton, can exist as 'superpositions' of many states at once. Yet
when any attempt is made to observe these states, we only
see one.

The question of why and how this is, is a central question in quantum mechanics, and has spawned many different proposals, or interpretations. The most popular is the
Copenhagen interpretation, which says nothing is real until it
is observed, or measured.

However, Copenhagen says nothing about what exactly constitutes an observation. John von Neumann broke this silence and suggested that observation is the action of a conscious mind. It's an idea also put forward by Max Planck, the founder of quantum theory, who said in 1931: 'I regard consciousness as fundamental. I regard matter as derivative from consciousness.'

That argument relies on the view that there is something special about consciousness, especially human consciousness. The conscious mind is somehow able to select out one of the quantum possibilities on offer, making it real – to that mind, at least.

Henry Stapp of the Lawrence Berkeley National Laboratory in California is one of the few physicists who subscribes to this notion. He says that we are 'participating observers' whose minds cause the collapse of superpositions. Before human consciousness appeared, there existed a multiverse of potential universes, Stapp says. The emergence of a conscious mind in one of these potential universes, ours, gives it a special status: reality.

There are many objectors. One problem is that many of the phenomena involved are poorly understood. Many philosophers, including Matthew Donald, a philosopher of physics at the University of Cambridge, argue that we don't even know whether consciousness exists, so making it a prerequisite for reality only adds to the confusion.

Donald prefers an interpretation that is arguably even more bizarre: 'many minds'. This idea – related to the 'many worlds' interpretation of quantum theory, which has each outcome of a quantum decision happen in a different universe – argues that an individual observing a quantum system sees all the many states, but each in a different mind. These minds all arise from the physical substance of the brain, and share a past and a future, but cannot communicate with each other about the present.

Though it sounds hard to swallow, this and other approaches to understanding the role of the mind in our perception of reality are increasingly being taken seriously. Understanding consciousness may well open a whole new philosophical can of worms.

Is consciousness a fourth state of matter?

'Solid, liquid, gas, mind: it's *all* about how you arrange the atoms,' says physicist Max Tegmark.

Imagine all the food you have eaten in your life and consider that you are simply some of that food, rearranged. This shows that your consciousness isn't simply due to the atoms you ate, but depends on the complex patterns into which these atoms are arranged.

If you can also imagine conscious entities, say aliens or future superintelligent robots, made out of different types of atoms then this suggests that consciousness is an 'emergent phenomenon' whose complex behaviour emerges from many simple interactions. In a similar spirit, generations of physicists and chemists have studied what happens when you group together vast numbers of atoms, finding that their collective behaviour depends on the patterns in which they are arranged. For instance, the key difference between a solid, a liquid and a gas lies not in the types of atoms, but in their arrangement. Boiling or freezing a liquid simply rearranges its atoms.

My hope is that we will ultimately be able to understand consciousness as yet another state of matter. Just as there are many types of liquids, there are many types of consciousness. However, this should not preclude us from identifying, quantifying, modelling and understanding the characteristic properties shared by all liquid forms of matter, or all conscious forms of matter. Take waves, for example, which are substrate-independent in the sense that they can occur in all liquids, regardless of their atomic composition. Like consciousness, waves are emergent phenomena in the sense that they take on a life of their own: a wave can traverse a lake while the individual water molecules merely bob up and down, and the motion of the wave can be described by a mathematical equation that doesn't care what the wave is made of.

If these efforts succeed, it will be important not only for neuroscience and psychology but also for fundamental physics, where many of our most glaring problems reflect our confusion about how to treat consciousness. There are promising prospects for grounding consciousness in fundamental physics. Much work remains, however, and the jury is still out on whether we will succeed.

4
Free will

When we make a decision – whether it's a simple one, like deciding to wriggle a finger, or a more complex one about whether to get married, we like to think that we are the ones in the driving seat. So how does this sit with what we know about the brain? And, more importantly, what has consciousness got to do with that process?

It is easy enough to spot actions that have nothing to do with free will. Reflexes, such as when you draw your hand away from a hot object, are a good example. You don't *decide* to move your hand, it moves anyway, triggered by your skin's contact with the hot object. The brain doesn't even get involved. There are many other kinds of actions, though, that have no clear external trigger, and seem to be generated entirely by the self. We can choose to act, and we can decide not to, the choice resides completely with us. Or so it seems. Yet when you start to look into what happens in the brain when we make the decision to do something, things get a whole lot less clear cut.

To study free will scientists use a kind of reverse engineering approach, starting with the action itself and then looking for the earliest signs of a decision. We have known since the work of neuroscientist Charles Sherrington, more than 100 years ago, that all of our voluntary physical movements come from the brain. Whenever we decide to move any of the muscles in our body, the movement is preceded by a particular set of brain activities. The only way to move your right hand, for example, is to first have activity in the left hemisphere motor cortex of the brain.

We can measure changes in brain activity over time using electroencephalography (EEG), a non-invasive brain-imaging technique that records tiny changes in the electrical signals passing through the brain's neurons, via electrodes on the scalp. This allows us to track brain activity before, during and after a decision has been made.

In 1965 neurophysiologists Hans Helmut Kornhuber and Lüder Deecke were the first to use this method to track activity in people who had been told to press a button whenever they liked. This experiment showed that the brain shows a gradual ramping-up of activity in the motor cortex about a second before a movement takes place. This they called the 'Bere-itschaftspotential', or 'readiness potential', because it seemed to represent the brain getting ready to press the button.

Who decides?

This finding raised the question of whether this readiness potential is part of conscious awareness. If not, does that mean that the decision to move happens before we know we are going it do it? And, if so, who is making that decision?

To answer this question we need to know the exact time that a conscious decision is made, and compare that to the timing of the readiness potential. An early attempt to tackle this question came from the work of Benjamin Libet, who did what became a classic experiment in the study of free will (*see* box).

The Libet experiments

In 1983, neuroscientist Benjamin Libet performed an experiment to test whether we have free will. Participants were asked to voluntarily flex a finger while watching a clock-face with a rotating dot. They had to note the position of the dot as soon as they became aware of their intention to act. As they were doing so, Libet recorded their brain activity via EEG electrodes attached to the scalp.

He, like Kornhuber and Deecke, found that a spike, 'the readiness potential', began up to a second before the movement itself. More importantly, it also began 350 milliseconds before the volunteers became consciously aware of their intention to act (*see* graphic below).

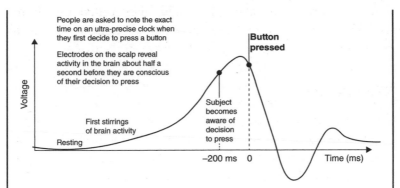

FIGURE 4.1 Who's in charge? An experiment that seems to challenge the notion of free will

Libet interpreted his results to mean that free will is an illusion. But we're not complete slaves to our neurons, he reasoned, as there was a 200-millisecond gap between conscious awareness of our intention and the initiation of movement. Libet argued that this was enough time to consciously veto the action, or exert our 'free won't'.

While Libet's interpretations have remained controversial, this hasn't stopped scientists carrying out variations of his experiment. Among other things, this has revealed that people with Tourette syndrome, who have uncontrollable tics, experience a shorter veto window than people without the condition, as do those with schizophrenia and healthy people who score highly on a standardized scale of impulsivity.

The surprising part of Libet's results is that there seems to be a period of time when the brain is preparing to do something, but you, yourself, don't yet know that you're going to do it.

This seems to conflict very strongly with our everyday notion that we – our conscious selves – decide what we are going to do and when.

To make sense of this, some psychologists point to a different view of voluntary action, in which it is unconscious brain activity that causes the body to move, but then, because you get sensory feedback about your own body movements, you can retrospectively insert your will back into the stream of consciousness. In this view, our sense of free will and our sense of having control of our actions before we make them is purely fictional. Like so much about our conscious experience, free will is just an illusion.

What if we don't have free will?

Our moral sense is based on an assumption so fundamental it seems unassailable: that we are masters of our own destiny. But what if neuroscience says otherwise?

One outcome may be a loosening of morals. Experiments have shown that people behave more selfishly and dishonestly if they are persuaded beforehand that free will is largely an illusion. They are also more likely to treat wrongdoers leniently, offering a hypothetical criminal a shorter prison sentence than they would otherwise have done. It's harder to ascribe blame to an automaton, after all. But these behavioural changes only last until the powerful feeling of our own agency reasserts itself.

On the other hand, belief in free will tends to be strengthened by considering a scenario in which someone acts immorally. Joshua Knobe of Yale University and his colleagues argue that our powerful belief in free will is bound up with a fundamental desire to hold others

---

responsible for their harmful actions. In other words, belief in free will is required to justify punishment. And there is some evidence that fear of punishment is what keeps societies from breaking down.

Without free will, would we reject punishment for crimes? One study suggests perhaps not. Experimental philosopher Eddy Nahmias at Georgia State University in Atlanta told 278 volunteers a story about a hypothetical future in which neuroimaging allowed for perfect predictions of decisions based on a person's brain activity. In this future world, a woman called Jill is fitted with a skull cap that allows scientists to predict everything she will do with 100 per cent accuracy. Yet 92 per cent of volunteers still thought that Jill's decision on who to vote for was her own free will. In another version of the story, the scientists didn't just predict the way Jill would vote, they also manipulated her choice via the skull cap. In that scenario most people said that Jill did not vote of her own free will.

It seems that even if neuroscience could show us that we aren't in the driving seats of our own minds, belief in our own free will is very tough to shake. This might be a good thing. Hanging on to a strong belief of our own agency is linked to a greater sense of satisfaction and self-efficacy, higher commitment in relationships and greater meaningfulness in life.

Perhaps, in practical terms, losing free will might matter less than we might fear. It may turn out that we don't have the ability to choose, but we will still choose to act as if we do.

FIGURE 4.2 A conscious choice or inevitable?

If the idea of free will being an illusion is unsettling, there is also some evidence to the contrary. This comes from experiments where electrodes are inserted into the brains of patients undergoing surgery for epilepsy.

During surgery, the neurosurgeon's aim is to target the source of the seizures without damaging neighbouring areas. To zero in on the correct part, the surgeon inserts grids of electrodes on the brain, and can stimulate these electrodes to identify what different brain regions do. Crucially, the patient is awake and fully conscious during the procedure, which means that certain parts of the brain can be stimulated and the patient can tell you exactly what they are experiencing.

In these kinds of experiments, neuroscientist Itzhak Fried stimulated a part of the brain called the **supplementary motor area**, which contributes to the body's physical movements. He found that when stimulated at low intensity, the patient reported an urge to move the limb that is controlled by the corresponding bit of brain, but the limb didn't actually move. Stimulate the same region at higher intensities, however, and the limb did move. This suggests that whatever is responsible for creating an urge to move is part of the same pathway that makes the movement actually happen.

What does all this mean for free will? One explanation is that, far from our urges and intentions being tricks of the mind, they are actually real mental states that stem from electrical activity in the brain. The question now is, does this 'gearing up to act' represent free will, or is it something else entirely?

The whole field of readiness potentials was thrown a curve ball in 2012 when Aaron Schurger, then of the National Institute of Health and Medical Research in Saclay, France, questioned the widely accepted idea that readiness potentials are the signature of the brain planning and preparing to move.

Previous research has shown that when volunteers were allowed to decide whether or not they pressed a button, the readiness potential was present regardless of their decision.

Schurger explained this by saying that readiness potentials aren't a specific kind of brain activity at all, but the result of random noise of the kind that bounces around the brain all the time. When we have to make a decision based on visual input, for example, groups of neurons start accumulating visual evidence in favour of the various possible outcomes. A decision is triggered when the evidence favouring one particular outcome becomes strong enough to tip its associated assembly of neurons across a threshold.

Schurger hypothesized that something similar happens in the brain during the Libet experiment. To find out, they repeated Libet's experiment, but this time if the volunteers heard a click while waiting to act spontaneously, they had to act immediately. The researchers predicted that the fastest response to the click would be seen in those in whom the accumulation of neural noise had neared the threshold – something that would show up in their EEG as a readiness potential. This is exactly what the team found: in those with slower responses to the click, the readiness potential was absent in the EEG recordings.

So while Libet argued that the readiness potential is the sign of an unconscious decision that is outside of free will, Schurger's work suggests that it is just the brain ticking along in no particular direction.

Since Schurger's original studies, further research has suggested that there is a measurable difference in the brain between a choice that is made deliberately (of our free will) or as a result of being told to do something.

So perhaps our voluntary actions don't just bubble up randomly like bubbles in boiling soup. We do have our own volition, and we can see it in the brain. However, since it's not part of consciousness at that stage, whether we can be held responsible for our actions is open to debate.

A sense of agency

The reason why free will is so hotly debated in both philosophy and neuroscience is because we all *feel* as if we have it. We have a clear sense of when something that happens in the environment was triggered by our own actions. How does this conviction happen in the brain?

The sense of agency is something that is difficult to study scientifically because we can't ever remember not having it. As babies we learn that when we bang a toy on the floor it makes a noise, and learn to put together a sequence of events which is initiated by something 'I' do. This very quickly gets wrapped up in our sense of self.

One way the brain produces the experience of control is by adjusting when we perceive our actions, and their outcomes. In one set of experiments, participants were asked to press a button, which caused a tone to occur 250 ms later. They reported either when they pressed the button, or when they heard the tone, using a rotating clock-hand, as in the Libet experiment (*see above*). They perceived actions that caused tones to occur later than actions in a control session which were not followed by tones. That is, the mind compresses the interval between action and outcome, highlighting the association between them. If the voluntary buttonpress is replaced by an involuntary movement, caused for example by stimulating the brain directly, this compression effect disappears, and is replaced by a repulsion effect, as if the brain were trying to separate in time the involuntary movement from the later tone.

As for the source of our sense of agency in the brain, two areas that seems to be particularly important are the **anterior insula** and the **angular gyrus**, which lies in the parietal cortex.

In one functional MRI study volunteers with joysticks moved images around on a computer screen. When the volunteers felt they had initiated the action, the brain's anterior insula was activated but the **right inferior parietal cortex** lit up when the volunteer attributed the action to the experimenter.

Interestingly, other researchers, using different experiments, have identified many more brain regions that seem to be responsible for the sense of agency.

Much is left to discover about the truth of free will and plenty of what we do know is open to debate. Nevertheless, neuroscientists now agree that when we make voluntary actions it's not because of any ghost in the machine, and it's not

FIGURE 4.3 Where there's a will...

because there's any sort of mind which is independent of the brain. Instead, there are brain processes which are just accompanied by a conscious experience, just as visual perception is a product of activity in the visual area of the brain. These specific brain processes give us our sense of control, our sense of being in charge of our body and our lives.

Without that, and without the feeling that you control your actions and, through them, the outside world, there would be no technology, no morality and, arguably, no human society.

The traditional philosophical question about free will is a question about where our actions come from. Thinking about the sense of agency inverts this question. Perhaps more important than where our actions come from is the ability to understand and represent what their consequences are. If you have learned the consequences of your own actions, you can at least pick up on learning signals in the environment which can tell you whether the outcome was positive or not, and thus whether you should make the action again.

An implant that channels our free will?

Imagine a world where you think of something and it happens. For instance, what if the moment you realize you want a cup of tea, the kettle starts boiling?

That reality is on the cards, now that a brain implant has been developed that can decode a person's intentions. It has already allowed a man paralysed from the neck down to control a robotic arm with unprecedented fluidity.

But the implications go far beyond prosthetics. By placing an implant in the area of the brain responsible for intentions, scientists are investigating whether brain activity can give away future decisions – before a person is even aware of making them. Such a result may even alter our understanding of free will.

The implant was designed by Richard Andersen at the California Institute of Technology in Pasadena for Erik Sorto, who was left unable to move his limbs after a spinal cord injury more than a decade ago. The idea was to give him the ability to move a standalone robotic arm by recording the activity in his posterior parietal cortex – a part of the brain used in planning movements.

Two tiny electrodes implanted in Sorto's posterior parietal cortex were able to record the activity of hundreds of individual neurons. After some training, a computer could match patterns of activity with Sorto's intended movement. Once this neuronal information had been collected, a computer translated Sorto's intentions into movements of a robotic arm. This enabled him to control the speed and trajectory of the arm, so he could shake hands with people, play rock, paper, scissors, and swig a beer at his own pace.

The breakthrough raises the tantalizing possibility of using other intentions decoded from brain activity to control our environment. For example, could we identify the pattern that corresponds to the thought of wanting to watch a film, then have that switch on the television?

To investigate the feasibility, Andersen's team had a person with a similar implant to Sorto's play a version of the prisoner's dilemma, where players can either collaborate or double-cross one another. The team was able to predict the volunteer's decision based on the neural activity the implant recorded. This showed that more abstract decisions such as, in this case, the intention to snitch on a hypothetical partner, can indeed be decoded from the posterior parietal cortex.

Eventually, he believes that a person with paralysis could imagine themselves making a cup of coffee and have a humanoid robot automatically carry out the action. He is hopeful that such approaches could one day be achieved using non-invasive techniques, such as recording brain activity with an EEG headset, rather than having to embed electrodes in the brain.

5
Disorders of consciousness

The wonders of consciousness should, by now, be abundantly clear. But the sad fact is that that anything that can go right can also go wrong. Research is offering new insights into disorders of consciousness, which may help treat some of the most distressing disorders of the mind.

The complexity of our conscious experience is such that disorders of consciousness come in many forms. In some cases the problem is one of altered level, or state of consciousness. Altered states can manifest themselves in sleep disorders, when the line between sleep and waking blurs, or in more serious cases when all control over consciousness is lost following brain injury. In others, they are due to a problem with organizing the contents of consciousness – our awareness, experience or memories. Sometimes the problem is one of binding these sensations and memories into a unified physical and mental sense of self.

Understanding what goes wrong in each of these cases will not only help us understand consciousness a little better, it might also help people escape from states of consciousness that none of us would choose.

Disorders of conscious level

While the content of someone else's consciousness can be difficult to ascertain from the outside, we generally reckon that we can make reasonably reliable judgements about another person's conscious state: a simple distinction between whether or not there is 'anybody home'. In medicine, doctors use the Glasgow Coma Scale to quantify consciousness, ascertaining the degree to which a patient is able to react and respond.

States of consciousness and the major disorders of consciousness can be mapped onto a graph that takes account of two dimensions: the presence or absence of behavioural signs of wakefulness and the presence or absence of awareness.

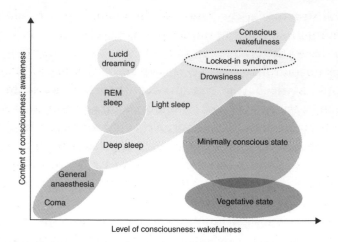

FIGURE 5.1 The many different levels of consciousness

Sleep disorders

Perhaps the best-known and most fascinating sleep disorder is **narcolepsy**, first described by the French doctor Jean-Baptiste-Édouard Gélineau at the end of the 19th century. Narcolepsy is a disorder of wakefulness characterized by two hallmark symptoms. The first is excessive daytime sleepiness, resulting in uncontrolled naps at the most unexpected times (during meals, exams, conversation, even sex). The second is **cataplexy**, a state most often brought about by laughter in which a person loses muscle tone and may collapse. Other symptoms include sleep paralysis, the disturbing experience of waking with the realization that you cannot move your body, and hypnogogic hallucinations, the vivid, dream-like imagery at sleep onset.

This strange constellation of symptoms appears to result from a lack of a neurotransmitter called **hypocretin**, discovered in 1998. This chemical, produced in the **hypothalamus,** a small structure at the centre of the brain, is sent to clusters of neurons that regulate the sleep/wake cycle. A lack of hypocretin leaves the brain in an unstable conscious state, in which it flips uncontrollably between sleep and wakefulness. In addition, 'REM sleep', the stage of sleep in which most dreams happen, is trigger-happy, coming on suddenly and unexpectedly throughout the day in people with narcolepsy. In REM sleep the body is normally paralysed to stop us acting out our dreams. Cataplexy is the paralysis of REM intruding into wakefulness, while sleep paralysis is the continuation of REM paralysis into wakefulness.

Several other sleep disorders can be understood in terms of abnormal overlap between the three fundamental states of consciousness: wakefulnesss, REM sleep and non-REM sleep (*see* Chapter 8 for more on sleep).

Loss of consciousness

Other disorders of conscious level can rob a person of human experience entirely.

Coma is a disorder that affects both wakefulness and awareness. A person in a coma has no sleep-wake cycle, shows no evidence of awareness of self or surroundings and, aside from reflexes, does not move. Comas can be caused by either a diffuse problem that affects both brain hemispheres – such as a traumatic brain injury – or a focal injury to part of the brain's central activating system, in the upper brain stem and thalamus.

The vegetative state is one of the outcomes of coma. This is a state of 'wakefulness without awareness' in which the sleep-wake cycle has recovered but there is no evidence of a functioning mind. The vegetative state can be eerie for loved ones since a person in this state can still respond to some degree,

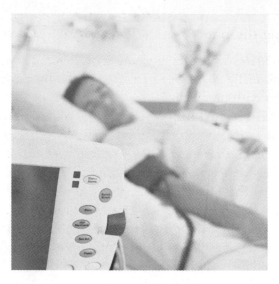

FIGURE 5.2 Coma: a disorder of wakefulness and awareness

for example by turning their head towards a sudden sound. They may also make emotional displays which are unconnected with what's happening around them. However, people who are genuinely in the vegetative state have very reduced metabolic rate in the brain, down to the kind of levels seen under general anaesthetic, and low activity in key cortical areas suggesting that for all intents and purposes there is indeed 'nobody home'.

Recovery from coma or vegetative state often involves a period spent in the 'minimally conscious state' with reproducible but inconsistent signs of awareness. Without very careful monitoring it can be very difficult to tell one from the other (*see* Show me you're there: ethical dilemmas of minimal consciousness, later in the chapter).

Zapping the brain from a minimally conscious state

In late 2016 something amazing happened. A man stuck in a minimally conscious state was roused by stimulating his brain with ultrasound.

The 25-year-old man, who had suffered a severe brain injury after a road traffic accident, progressed from having only a fleeting awareness of the outside world to being able to answer questions and attempt to walk.

He was the first person in the world to undergo an experimental procedure by Martin Monti at the University of California, Los Angeles, which stimulates the thalamus deep inside the brain using pulses of ultrasound. Monti and his team have been searching for a way to help people with brain injuries that result in disorders of consciousness, for whom there are few treatments. In a minimally conscious state a person shows fluctuating signs of awareness, but cannot communicate.

Animal experiments suggested that stimulating the thalamus might help promote arousal. The thalamus is a kind of hub situated in the centre of the brain, which acts as a relay between many other regions. Rats under anaesthesia can be woken faster after thalamic stimulation, for example. In 2007, a 38-year-old man who had been in a minimally conscious state for six years showed some signs of recovery after receiving deep brain stimulation to his thalamus, but this involved implanting electrodes into his brain, a procedure that risks damaging other areas.

Monti's team decided to try using low intensity ultrasound instead, which can safely modulate deep brain tissues without harming the surrounding areas.

They tried this technique on a patient who had been in a minimally conscious state for almost three weeks. Monti placed an ultrasound transducer on the patient's temple, which is directly above the thalamus, seven centimetres below. They then stimulated the man's thalamus for 10 minutes, alternating between 30 seconds of ultrasound and 30 seconds rest.

The following morning the patient was starting to vocalize and gesture in response to questions, behaviours he had not shown before treatment. Over the next three days the man started answering questions by nodding or shaking his head. He even managed to fist-bump Monti. A week later, the patient tried to walk.

It's certainly exciting, but since it has only been done in one patient, it is too early to conclusively say that it was the treatment that made the difference. The patient was young, which makes it possible that his brain may have recovered with or without help.

The team now hopes to test the procedure on 10-15 more people with disorders of consciousness, and to trial it in healthy people, to see the effects of stimulating and inhibiting a normal thalamus.

It is not yet clear whether the thalamus underlies the fundamental aspects of conscious awareness, or if it is instead involved in helping patients produce behaviours that give a sign that they are aware of the world around them. Further studies may solve the mystery.

Locked in

The most terrifying disorder of consciousness to many people is the **locked-in state**, where a person is fully conscious but unable to communicate that awareness to the outside world. This, technically, is not a disorder of consciousness at all, since the sufferer is very much aware, but from the outside it can be hard to distinguish from coma. It classically occurs when a stroke in the brain stem prevents the normal control of limbs and voice box: in such cases the clue to the diagnosis is that the patient can typically communicate by looking up or down or opening and closing their eyes, as these functions are controlled above the level of the stroke. However, it is possible, in principle, to be wholly paralysed – for example by a paralysing drug – and yet be fully aware.

Now, thankfully, there are ways of peering into the brain to measure the level of awareness and wakefulness when people are unable to report it for themselves. Around ten years ago, Adrian Owen, a British neuroscientist now at the University of Western Ontario in Canada, found that healthy people produce a particular pattern of brain activation when you ask them to imagine playing tennis, which is different to the pattern of activation when you ask them to imagine walking around their house. In a group of about 20 people who appeared to be vegetative, Owen found one patient who showed the same pattern of activation as healthy volunteers and concluded that she must be aware. Indeed, she started to show signs of awareness some weeks later.

It's since been found that about 10 to 20 per cent of people who appear to be in the vegetative state show this pattern of activity when they are given the opportunity, suggesting that the vegetative state can easily be misdiagnosed.

Another measure is the Perturbational Complexity Index (PCI) described in detail in Chapter 2. The PCI is reduced in sleep, under anaesthesia and in the vegetative state, with an increase in the minimally conscious state, and shows a normal value in the locked-in state. Using these techniques it is becoming easier to tell one disorder of consciousness from another and treat patients accordingly.

Disorders of content

If it is difficult to get a handle on the level of someone else's consciousness, it is even trickier to access the contents of that inner life. This is partly because it is difficult to make objective measurements of what is going through someone else's mind, but also because the contents of our experience are constantly shifting – for example, from the present moment to memories of our past, and our plans for the future.

We do know, however, that problems with the brain's sensory and memory apparatus can lead to problems with particular kinds of conscious content.

In a rare condition called **hemichromatopsia**, for example, a person sees one half of their visual world as black and white. This happens because of damage to an area of the visual cortex, called V4, which adds the experience of colour to incoming information from the primary visual cortex.

The brain has other areas that specialize in particular kinds of visual processing; damage to these can selectively knock out content. The **fusiform face area**, for example, is activated when you see or think about the face of someone you know. Damage to this area leaves people with a condition called **prosopagnosia**, in which they are unable to recognize people by their faces alone – even sometimes their own face in the mirror. Research has shown that some people are born with prosopagnosia, but often don't realize this because they recognize people by the way they walk, talk or do their hair. Knocking out one element of conscious content forces the brain to compensate with other inputs.

Problems with brain regions that deal with memory can have serious implications for our conscious awareness. Henry Molaison, the famous American patient known until his death in 2008 by his initials, H.M., had both **hippocampi** removed as a young man to cure his epilepsy. As a result he lost the ability to lay down autobiographical memories and to think about the future, limiting his conscious experience to the here and now for the rest of his life. Epilepsy in brain areas adjacent to the hippocampus can cause **déjà vu**; a disorder of the timeline of our conscious experience or a glitch in the matrix of our minds.

Alzheimer's disease, too, begins in the hippocampus and surrounding areas, and the loss of memory that comes as an early sign of the disease hints at the loss of self that will follow. The hippocampus is highly linked to a set of structures in the brain that together make up what is known as the **default mode network**. These are the areas of the brain that are particularly active when we are 'at rest', without any particular task to perform, and are involved in recollection of the past, anticipation of the future and understanding other people's minds – in other words, the contents of our inner lives. This network is also affected in early Alzheimer's, giving rise to the involvement of memory typical of this disease.

Other dementias affect other brain regions, and consequently different aspects of experience. **Semantic dementia**, for example, affects the **lateral temporal neocortex**, roughly behind the ears, which is where our database of knowledge about language and the world is held. Damage here leads to word-finding difficulties and loss of knowledge, while sparing day-to-day memory of the kind that is especially affected early in Alzheimer's disease.

Disorders of imagination

Most of what we know about disorders of consciousness concerns problems that affect moment to moment awareness of
the world, but that isn't the only way in which things can go
awry. In recent years we have begun to identify disorders of
our 'extended consciousness'. Our human awareness is greatly
enriched by our ability to detach ourselves from the here and
now, to recollect the past, to travel imaginatively into the future,
and to enter the virtual world of a novel or scientific theory. We
spend a great deal of our human lives in this altered state and
this too, it seems, can go wrong.

Around ten years ago, a man, known as MX in scientific studies, lost his ability to visualize following surgery for a blocked
artery (*see* box: Point of view: What's it like to lose your mind's
eye?). Functional brain imaging of MX's brain while he was
either looking at a face or trying to visualize it showed that
while in most people the same brain areas are activated when
seeing and imagining, in this man the corresponding brain areas
failed to activate when he tried to visualize.

Point of view: What's it like to lose your mind's eye?

*After surgery, a man known in scientific papers by the initials,
MX, lost his ability for visual imagery. Here, he tells his story to
Adam Zeman.*

MX: I found that I first noticed it just a few weeks after I
had my angioplasty. I noticed at night when I went to bed
that I couldn't do what I usually did, which was, before
going to sleep, thinking about my family, my children and

grandchildren and picturing them. They just wouldn't come to mind. And also I used to have a technique to go to sleep if I couldn't get to sleep, and that was to start at 99 and count backwards by looking at the numbers, either black on white, or white on black, and see them in my mind's eye and click them off. I never got down to one, I was asleep by then. So that was about three weeks after I'd had the angioplasty.

AZ: Were you able to visualize places that you'd visited?

MX: No. I couldn't visualize anything. I could remember them but I couldn't visualize them.

AZ: Has it affected your memory?

MX: As far as I can see it hasn't had any effect at all. My memory is as good as it was, I think.

AZ: Even to remember visual details?

MX: Yes. I know that sounds a bit odd, even to remember visual details which I can't see, but I can remember. I just know, but I can't see them. I can't explain that.

The description of this case led to widespread publicity and the recognition that substantial numbers of people – perhaps 2 per cent of the population – lack this ability. This has been described as lifelong **aphantasia** ('phantasia' was Aristotle's term for the mind's eye, the 'a' denoting its absence).

Some people who report a lifelong inability to visualize are affected across all the senses – so have no 'mind's ear' or 'mind's tongue' – while about half of affected people lack only visual imagery. Some people with this problem describe difficulty with autobiographical recollection, which perhaps shouldn't be surprising because visualization is for most of us a very important ingredient in autobiographical recollection. Indeed, many of the areas in the brain which are activated when we recollect our past are also activated when we visualize. Nevertheless, some artists and novelists report that they lack imagery, but are highly creative, so clearly lack of visualization is not the same as lack of imagination. Craig Venter, who was the first person to decode the human genome, has long recognized that he lacks visual imagery and claims that this has contributed to the strength of his work.

Interestingly, many people with aphantasia know what imagery is like since they dream visually, and experience hypnogogic images as they drift off to sleep. However, they can't summon up imagery voluntarily. This may be because dreaming is essentially a bottom-up process. Activity in the brain stem drives dreaming, and brain activation when we dream is very different to brain activation when we are awake and alert. When we make a voluntary decision to visualize, we are operating 'top-down' and use a very different set of brain structures and networks to drive the process, particularly involving regions of the frontal and parietal lobes.

At the other end of the scale, another group of people has recently been identified that has unusually vivid mental imagery – **hyperphantasia**. Comparing these two groups of people should shed new light on the way we construct our internal reality.

Interview: Show me you're there: Ethical dilemmas of minimal consciousness

What if some people who seem to be in a vegetative state are actually conscious, asks ethicist Joseph Fins, professor of medical ethics and medicine at Weill Cornell Medical College, New York, and co-director of CASBI, the Consortium for the Advanced Study of Brain Injury.

Your work concerns the tough questions raised when brain injuries affect consciousness. Why is this issue so important to you?

We have only recently come to realize that a subset of people diagnosed as being in a vegetative state were not vegetative at all. They were not permanently unconscious; in fact, they were conscious. We now know this as the 'minimally conscious state'. It struck me as a human rights issue, that these people who had the ability to interact and be aware at some level were sequestered in nursing homes.

So in your book, *Rights Come to Mind*, you decided to tell their stories

I interviewed more than 50 families of people who had come to Cornell and Rockefeller for our studies on how the brain recovers from disorders of consciousness. I felt a tremendous moral obligation to give a voice to these voiceless people and their families who were struggling, often in isolation and overwhelmed by grief.

You illustrate these dilemmas with the story of Maggie Worthen. Tell me about her

Maggie was a senior in college when a brainstem stroke left her in what was thought to be a permanent vegetative state. Two years later, her mother sought us out, wanting to find out if Maggie had some awareness. We were able to demonstrate behaviourally, and then with neuroimaging, that she was indeed minimally conscious.

One time, my colleague Nicholas Schiff pointed to Maggie's mother and asked 'Is that your mother?' There was this long pause, and then this downward swoop of Maggie's eye – which meant 'yes'. Then Maggie's mother started sobbing on my shoulder. That was a pivotal moment.

We've known about the minimally conscious state since 2002. Why is misdiagnosis still a problem?

The challenge with the minimally conscious state is that the behavioural manifestations are intermittent, so cannot always be reproduced. That makes it complicated to identify and the misdiagnosis rate is high. One study showed that 41 per cent of people with traumatic brain injuries in nursing homes who were diagnosed as vegetative were actually minimally conscious.

I argue that we have to prepare society for the consequences of advances in neuroscience which will expose us to new problems, but also give us opportunities to find new solutions.

What kinds of scientific advances are changing the picture?

Neuroimaging can help us find out if these people are responsive and conscious. Neuroscience has made us aware of this situation, and may also help to deal with it, through neuroprosthetics, deep brain stimulation, other devices and drugs. But, fundamentally, it's a societal issue because now we have people we treat who deserve more than we have traditionally given them.

What was it like being part of the first team to try deep brain stimulation to help improve the consciousness of people with brain injuries?

That work started through my collaboration with Nicholas Schiff, who is at Weill Cornell with me. We were interested in the disconnect between what you see overtly and what is going on internally. He had this idea of using deep brain stimulation to help restore functional communication in the minimally conscious state. That presented a huge ethical challenge: how do you do research on someone who can't give consent? I spent the better part of ten years working on the ethical formulation that would make such work possible. That project was published in *Nature* in 2007. We found that deep brain stimulation could lead to improvements in attention, limb control and spoken language. A subject in the trial was in a minimally conscious state after being assaulted, and through this procedure gained enough coordination to eat by mouth.

You talk about these issues in a very hopeful way, but many of the experiences of people you describe are pretty awful

I'm not trying to romanticize these brain states. Nobody would choose to be this way. But after a brain injury, families first hope their loved one will survive and wake up. Then they hope that when they wake up they're conscious.

Then they hope that when they're conscious they're more than minimally conscious – but then they end up in a place they didn't necessarily expect to reach. We're trying to help these people regain as much functional capacity as they can. A recent study showed that 22 per cent of people with a disorder of consciousness will recover enough to live independently. Most people are staggered when they hear that number.

What is the paradox you talk about between the care people get at the start of treatment and later on?

People who have traumatic brain injuries initially receive brilliant medical care that saves their life. But after a traumatic brain injury, people often end up in chronic care facilities because they're not considered ready for rehabilitation. But once there, they often don't come back.

So you're saying the chance to improve their condition is missed?

Yes. Somebody may be in the vegetative state when they're discharged from hospital. They might end up in a nursing home and then start exhibiting responsive behaviours that are intermittent; a doctor is called but they are not reproduced. If the doctor doesn't know about this new science, they might ascribe it to family denial when it's actually the biology of the minimally conscious state.

I describe cases like this in the book. Terry Wallis, for example, started talking 19 years after his injury. His family thought they saw things throughout that time, but it wasn't until he emerged from the minimally conscious state and started talking that people appreciated he had been minimally conscious for most of those 19 years.

Can people who recover help change perceptions of what it's like to live with these kinds of brain injury?

The problem is that the people who were minimally conscious don't remember that time. Everybody wants to know what they were thinking. But they don't remember because the hippocampus, where memory resides, is one of the most exquisitely sensitive parts of the brain to injury and other kinds of trauma – so they don't have a recollection. But their stories are becoming exemplars of why we should be worried.

6
Conscious machines

When so much is left to understand about human consciousness, what are the chances of building something similar into robots?

Building 'feel' into machines

We can learn a lot about consciousness by thinking about what is would mean for a robot to be aware of itself, to feel and to think.

In 2016 a team of German researchers claimed that they had made a robotic arm that could feel pain. Sure enough, when the robot was tapped hard it would react more strongly, pulling itself away more quickly than if it were touched gently. But is that the same as feeling pain? Most people would probably say not: they would say that the robot has been programmed to react in different ways to different kinds of stimuli, but that it doesn't actually *feel* anything.

As human beings we don't just sense and react, we really *feel*; experiencing rich and vivid sensations that are more than just reactions. Science has been unable to come up with a convincing explanation of why sensations feel the way they do. Some people even insist that understanding the nature of sensations is outside the realm of science. Given this, it's tempting to think that building anything like sensations into a robot should be impossible.

The most influential ideas on how to account for 'qualia', that is, the particular qualities of experience, suggest that some poorly understood and complex process in the brain might cause different feels to emerge. But this approach raises more questions than it answers. Imagine we had discovered that the difference between the experience of red and the experience of green was due to some brain process, say the different frequencies of oscillation in the visual cortex, for example. Then you could ask, *what is it* about the different frequencies that gives you the red feel rather than the green feel? The puzzle would continue to exist. Indeed, no matter what neural explanation is put forward to account for the difference between red and green, you could always ask a similar question: What is it about that particular brain mechanism that gives you the experience of red rather than green? The reason for the puzzle is that there seems to be no common language that links the quality of experience, which we can't even describe in words, and possible physical or neurological mechanisms.

Thinking about 'feel'

Perhaps, then, it is time to think differently about 'feel'. The sensorimotor theory of consciousness, (*see* Chapter 2) describes the experience of 'feel' not as a mysterious by-product of the brain, but as *the way we interact with the world around us.*

For example, imagine pressing on a sponge. Where is the feel of softness generated? The traditional approach would look for the sensation in the brain, but press down on the nearest soft object right now and it quickly becomes apparent that the softness isn't in the brain at all: it resides in the particular characteristics of the action involved in pressing a soft object. Sensorimotor theory proposes that all 'feels' might be like this. The colour of red, the smell of onion and the sound of a bell, are all different ways of interacting with the world. If this is so, then feel is certainly something that could be experienced by a robot – it is certainly easier to achieve than sensations that mysteriously emerge from the complex activity of the brain.

Sensorimotor theory takes the mystery out of 'qualia'. It explains what red is like, what vision is like, what hearing is like, in the same words that we use to describe what we do when we interact with the world. Crucially, robots should also be able to interact in such ways.

So what factors are required to generate human-like experiences? One important feature of our sensations is that they have 'sensory presence': they seem to come from outside of us and have a real, physical presence compared to other things that happen in the nervous system like thinking, imagining or remembering.

Bodiliness

One reason for this sensory presence of our sensations is 'bodiliness'. Sensory input coming from one of the five sense modalities depends very strongly on bodily movements. If you are looking at something and move your eyes, or body, there's an immediate change in information coming into your nervous system. Compare that with thinking: if you're thinking about something and move your body, there's no change in what's coming in. Thoughts have no 'bodiliness', and so are not experienced as coming from the outside world.

In addition to our bodies, the outside world also has an active role to play in our sensory experiences: inputs to the nervous system are not just affected by our body motions, but also by changes in the environment around us. In sensorimotor theory this is called **insubordinateness** to the outside world, since the outside world can butt in and take centre stage at any moment.

This hints at another way in which sensory inputs are different from other brain activities like thought or memories. Sensory inputs are 'grabby': when they change abruptly, they immediately capture our attention. This 'grabbiness' comes from the fact that evolution has wired up humans' sensory systems so that ongoing cognitive processing can be immediately interrupted by a sudden environmental event.

FIGURE 6.1 How could we ever be sure that an android is conscious?

So, if we were to build bodiliness, insubordinateness and grabbiness into a robot's sensory systems, would it be conscious? Not necessarily. To consciously experience a feel, an agent has to not only sense it, but it has to pay attention to what it is sensing. As an example, you can drive your car while talking to the passenger, stopping at the red light and taking all the correct turns, but when you get home you might not remember any of the detail. So being conscious of something also requires you to be paying attention.

Do we need hardware to solve the hard problem?

Most computers and robots created so far run on software. This, says Pentti Haikonen, an electrical engineer and philosopher at the University of Illinois in Springfield, is why they will never experience pain or colours in the same way that we do. Software is a language, he says, and so requires extra information to be interpreted. If you don't speak English, the words 'pain' or 'red', for instance, are meaningless. But if you see the colour red, that has meaning no matter what your language.

Haikonen has made a robot, called XCR or eXperimental Cognitive Robot, that stores and manipulates incoming sensory information, not via software, but through physical objects – in this case wires, resistors and diodes. Sensations like 'red' and 'pain' are direct experiences to the brain, with no interpretation on the way.

'XCR has been built so that if hit with sufficient force the resulting electrical signal makes it reverse direction – an avoidance response corresponding to pain,' Haikonen says. The robot is also capable of a primitive kind of learning. If, when it is hit, it is holding a blue object, say, the signal from its blue-detecting photodiode permanently opens a switch. From now on, the robot associates the colour blue with pain and reverses away uttering 'Me hurt, blue bad.' Try to push it towards an object again and it backs away. 'Blue, bad.'

As robot achievements go, learning to avoid a blue object is no big deal: conventional software-based robots can do it standing on their heads. But the fact that XCR bypasses software, storing sensory information directly in its hardware, takes it the first step down the road to awareness, claims Haikonen.

If he is right, and we can't create a feeling machine based on software, then no matter how big the net gets, it will never be sentient. But a brain in a vat wired up to a supercomputer simulation – a classic thought experiment from philosophy – could be conscious.

So if we were to build an attentional processor into the robot, now can we say it can feel like we do? Again, there's something missing. Healthy humans have selves: there is an 'I' doing the feeling. If a robot were able to pay attention to what needs to be processed, yet didn't know that it itself existed, it surely would not be experiencing anything.

Now, building a sense of self into a robot may actually not be as unattainable as you might think. There are many levels of self-knowledge. An amoeba is able to avoid eating itself, so at a basic level it is able to distinguish itself from other things. At a higher level, a squirrel will hide its nuts so that other squirrels can't get them. This shows that the squirrel distinguishes itself from other squirrels in terms of its own goals. And then humans – perhaps dogs, dolphins, primates – have a higher level of self-knowledge which you could call knowledge of self-knowledge, which gives rise to more complicated phenomena like empathy, beliefs, desires and motivations. Building these into a robot is not as impossible as it might seem (*see* box below).

Nao, the self-aware robot

In a robotics lab on the eastern bank of the Hudson River, New York, in 2015, three small humanoid robots were given a conundrum to solve.

They are told that two of them have been given a 'dumbing pill' that stops them talking. In reality the push of a button has silenced them, but none of them knows which one is still able to speak. That's the problem they have to work out.

Unable to solve it, the robots all attempt to say 'I don't know'. But only one of them makes any noise. Hearing its own robotic voice, it understands that it cannot have been

---

silenced. 'Sorry, I know now! I was able to prove that I was not given a dumbing pill,' it says. It then writes a formal mathematical proof and saves it to its memory to prove it has understood.

This is the first time a robot has passed a classic test called the wise-men puzzle. It sounds like a simple test and it is, hardly scaling the foothills of consciousness. But showing that robots – in this case, off-the-shelf Nao models – can tackle logical puzzles requiring an element of self-awareness is an important step towards building machines that understand their place in the world.

Selmer Bringsjord of Rensselaer Polytechnic Institute in New York, who ran the test, says that by passing many tests of this kind – however narrow – robots will build up a repertoire of abilities that start to become useful. Instead of agonizing over whether machines can ever be conscious like humans, he aims to demonstrate specific, limited examples of consciousness.

The wise-men test requires some very human traits. The robots must be able to listen to and understand the question 'which pill did you receive?', as asked by a human. They must then hear their own voice saying 'I don't know' and understand that it was they that said it, connecting that with the idea that they did not receive a silencing pill.

Bringsjord's robots may appear conscious in this specific case, assessing their own state and coming to a conclusion. But the broader, deeper intelligence that we humans have is completely missing. The Nao robots can pass the wiseman test but wouldn't have a hope of recognizing their own feet.

Bringsjord says one reason why robots can't have broader consciousness is that they just can't crunch enough data. Even though cameras can capture more data about a scene than the human eye, roboticists are at a loss as to how to stitch all that information together to build a cohesive picture of the world.

The test also shines light on what it means for humans to be conscious. What robots can never have, which humans have, argues Bringsjord, is phenomenological consciousness: 'the first-hand experience of conscious thought', as Justin Hart of the University of British Columbia in Vancouver, Canada, puts it. It represents the subtle difference between actually experiencing a sunrise and merely having visual cortex neurons firing in a way that represents a sunrise. Without it, robots are mere 'philosophical zombies', capable of emulating consciousness but never truly possessing it.

Even with all of these qualities, though, a smart, sensing robot with self-awareness would only be truly conscious if it could persuade the rest of us of its own inner experience. Even in humans, the notion of a self can be considered to be a cognitive or social construct. It's what the philosopher Daniel Dennett calls a **narrative fiction**: we are telling a story about our own inner lives and those of others. Why does it feel so real? Consider other kinds of cultural constructs, like money. Money is a story in the sense that we all agree about the value of certain pieces of paper and certain bits of metal. If we didn't agree that they had value, they would be worthless. The self is the same: though it is a story, it is perfectly real.

This same could be said of a sensing, self-aware robot. As they become cleverer and as they become more and more

FIGURE 6.2 Who are you calling a zombie?

integrated into our society – and perhaps into their own – they too will say of us and of themselves that they are conscious. And at that point they will feel as we feel. Of course there will be differences, because they have different bodies, different modes of interaction, but their 'feels' will be as real to them as ours are to ourselves. Feeling robots will live among us, perhaps sooner than we think.

You, in silico

Imagine never having to truly die. Being able to upload all of your memories to a computer, and to live on in a humanoid robot when your human body can no longer be upgraded.

It is a fantasy that is still a long way from reality but several companies are taking the first steps in that direction.

The initial goal is to enable you to create a lifelike digital representation, or avatar, that can continue long after your biological body has decomposed. This digitized 'twin' might be able to provide valuable lessons for your great-grandchildren – as well as giving them a good idea of what their ancestor was like.

Ultimately, however, they aim to create a personalized, conscious avatar embodied in a robot – effectively enabling you, or some semblance of you, to achieve immortality.

So far, the available options are little more than a glorified social media presence, with an artificially animated photo, but several companies are working on not only perfect reproductions of faces, but on capturing unique facial expressions and natural speech.

Many challenges remain, not least the time and expense involved in creating a realistic-looking avatar that knows all about your personality and tastes. To create a truly life-like representation would take a lifetime of training, and let's face it, most of us are too busy having a life. But in the future, who knows.

The agricultural revolution saw a vast expansion of what human beings could accomplish together while the industrial revolution saw power shift from rural nobility to urban business. The digital identity revolution in its turn could transform how people think about themselves, their life, and what it means to be human.

Watch this space.

7
Altered states

What can strange states – from out-of-body experiences to blissed-out hallucinations – tell us about how consciousness works?

Going under: how hypnosis alters our awareness

Hypnosis has something of a shady reputation among scientists, but that hasn't stopped a handful of researchers from studying it as a possible window into the nature of consciousness.

David Oakley, an emeritus professor at University College London, is using it to induce unusual states of mind in otherwise healthy people. The idea is to create 'virtual patients' with symptoms that can literally be switched on and off with a snap of the fingers, making it easier to study the abnormal brain activity that causes them.

Oakley, along with Peter Halligan, a neuropsychologist at Cardiff University, has focused on a range of rare and bizarre conditions in which normal conscious awareness is disrupted.

They include **hysterical blindness,** in which a person cannot consciously see but has no perceptible damage to their eyes or brain, and **visual neglect** in which a person lacks awareness of half of the visual field. Also, **hysterical paralysis,** an inability to move a part of the body despite having no physical injury is a disorder of free will in which voluntary movement is interrupted. The same limb may move with no problems while the person is asleep. **Alien limb syndrome** is another problem with free will in which it feels as if an arm or leg is acting of its own accord (*see* box below).

Free won't: when a limb has a life of its own

People with so-called **anarchic hand syndrome** find that their affected limb reaches out and grabs things they have no wish to pick up. They might try restraining it with their other hand, and if that doesn't work, they sometimes tie it down.

The cause is injury to the brain, usually in a region known as the **supplementary motor area** (SMA). Work on monkeys has shown that another part of the brain, the **premotor cortex**, generates some of our actions unconsciously in response to things we see around us. The SMA then kicks in to allow the movement or stop it, but damage to the SMA can wreck this control – hence the anarchic hand, acting on every visual cue.

A few people are unfortunate enough to have damage to the SMA on both sides of the brain, and experience both hands acting outside their control. They are at the mercy of environmental triggers, says Sergio Della Sala, a neuroscientist at the University of Edinburgh, UK, who studies the condition.

The system sounds like the very opposite of free will – Della Sala calls it 'free won't'. The findings suggest that, while it feels like our actions are always under our conscious control, in fact there is a lot of unconscious decision-making going on too.

Oakley and Halligan believe that these conditions can be recreated in healthy people using hypnosis, potentially shedding light on what causes them. This, they hope, will fast forward our understanding of conditions that are hard to study since they are both incredibly rare and often found in people with other problems, such as depression or schizophrenia.

Can the activity in a hypnotized brain really mimic what's going on in someone with a real disorder? Oakley and Halligan are convinced their virtual patients are experiencing some of the same brain changes as people with genuine disorders.

Halligan tells how they once induced a case of visual neglect in a volunteer by suggesting that the left side of his visual field would cease to exist. They then asked him to copy a picture with a dozen objects scattered on the page. Most hypnotized people given this instruction would copy only the objects on the right-hand side of the page, as most people with visual neglect do. But this volunteer, like real patients, drew the right-hand side of every object on the page.

The similarities reach into the brain, he says. Halligan and colleagues put 12 highly hypnotizable students under and then either suggested that their left leg was paralysed, or told them to merely pretend that their left leg was paralysed, with the promise of a reward if they managed to fool an investigator. The investigators, unaware of which group the participants were in, couldn't tell who was faking paralysis – until they saw scans of the volunteers' brains. There were clear differences in brain activity. One of the brain areas that was highly active, or 'lit up', in the hypnotically paralysed volunteers was the **right orbitofrontal cortex** – a region thought to be involved with emotional inhibition, and which has also been seen to be active in hysterical paralysis.

This work suggests that brain areas normally associated with the intentional inhibition of movement are not active in people with hysterical paralysis nor hypnotized volunteers. This suggests that it really is a problem with free will. It is not that they will not move, but they *cannot* make it happen.

Whether such studies will help to develop treatments for patients remains to be seen. Perhaps the biggest impact will simply be to convince people that the conditions are real. Being able to show both doctors and patients that these strange states of awareness are real conditions is a big step forward.

Not going under

If hypnosis leaves you unmoved, blame the wiring in your brain. It seems those who find it easier to fall into a trance are more likely to have an imbalance in the efficiency of their brain's two hemispheres.

Around 15 per cent of people are thought to be extremely susceptible to hypnosis, while another 10 per cent are almost impossible to hypnotize. The rest of us fall somewhere in between. Sceptics argue that rather than being in a genuine trance, some of us are simply more suggestible and therefore more likely to act the part. However, recent studies have hinted that during hypnosis, there is less connectivity between different regions, less activity in the left side of the brain and more in the right side. Such findings suggest hypnosis is more than acting.

To see if there are also differences between the brains of susceptible and unresponsive volunteers when they were awake, Peter Naish of the Open University in Milton Keynes, UK, used a standard test of hypnotic susceptibility that combines motor and cognitive tasks to identify ten volunteers of each type. He then gave each volunteer a pair of spectacles with an LED mounted on the left and right side of the frame. The two LEDs flashed in quick succession, and the volunteers had to say which flashed first. Naish repeated the task until the gap between the flashes was so short that the volunteers could no longer judge the correct order.

Naish found that hypnotically susceptible volunteers were better at perceiving when the right LED flashed first than when the left one did. This suggested that the left hemisphere of their brain was working more efficiently (visual pathways cross over in the brain, so left controls right and vice versa). In contrast, the non-susceptible people were just as likely to perceive the right LED flashing first as the one on the left.

These differences in the balance of brain efficiency persisted when Naish tried to hypnotize both groups. During hypnosis, the brains of those in the susceptible group seemed to switch 'states', becoming faster at spotting when the left LED flashed first. Meanwhile, the efficiency of the hemispheres remained relatively even in the non-susceptible people. They didn't fall into a trance, but their performance on the task started to deteriorate.

Naish suggests that successful hypnosis requires temporary domination by the brain's right side, a state that might be much easier to bring about in people who tend to have an imbalance in the efficiency of their two hemispheres, even when awake. This fits in with a theory that hypnosis involves a transition from left to right hemispheric dominance. Zoltan Dienes of the University of Sussex in Brighton, UK, has used **transcranial magnetic stimulation** to temporarily reduce activity in the left hemisphere and found that this increases responsiveness to hypnosis, perhaps giving the volunteers a helping hand by reducing activity in their left hemisphere.

Out-of-body experiences

One morning a young man woke feeling dizzy. He got up and turned around, only to see himself still lying in bed. He shouted at his sleeping body, shook it and jumped on it. The next thing he knew he was lying down again, but now seeing himself standing by the bed and shaking his sleeping body. Stricken with fear, he jumped out of the window. His room was on the third floor. He was found later, badly injured.

What this 21-year-old had just experienced was an out-of-body experience, one of the most peculiar states of consciousness. It was probably triggered by his epilepsy. He later told doctors that he wasn't trying to kill himself. He jumped in a desperate attempt to rejoin body and self.

Since that dramatic incident scientists, including Peter Brugger, the young man's neuropsychologist at University Hospital

FIGURE 7.1 Out-of-body experiences are one of the most peculiar states of consciousness

Zurich in Switzerland, have come a long way towards under-standing out-of-body experiences. They have narrowed down the cause to malfunctions in a specific brain area and are now working out how these lead to the almost supernatural expe-rience of leaving your own body and observing it from afar. They are also using out-of-body experiences to tackle a long-standing problem: how we create and maintain a sense of self.

Dramatized to great effect by such authors as Dostoevsky, Wilde, de Maupassant and Poe – some of whom wrote from first-hand knowledge – out-of-body experiences are usually associated with epilepsy, migraines, strokes, brain tumours, drug use and even near-death experiences. It is clear, though, that people with no obvious neurological disorders can have an out-of-body experience. By some estimates, about 5 per cent of healthy people have one at some point in their lives.

Your doppelgänger

So what exactly is an out-of-body experience? A definition has recently emerged that involves a set of increasingly bizarre perceptions. The least severe of these is a doppelgänger experi-ence: you sense the presence of or see a person you know to be yourself, though you remain rooted in your own body. This often progresses to stage 2, where your sense of self moves back and forth between your real body and your doppelgänger. This was what Brugger's patient experienced. Finally, your self leaves your body altogether and observes it from outside, often from an elevated position such as the ceiling.

Some out-of-body experiences involve just one of these stages; some all three, in progression. Bizarrely, many people who have one report it as a pleasant experience. So what could be going on in the brain to create such a seemingly impossible sensation?

The first substantial clues came in 2002, when Olaf Blanke, a neurologist at the Swiss Federal Institute of Technology in Lausanne, and his team stumbled across a way to induce a full-blown out-of-body experience. They were performing exploratory brain surgery on a 43-year-old woman with severe epilepsy to determine which part of her brain to remove in order to cure her. When they stimulated a region near the back of the brain called the **temporoparietal junction** (TPJ), the woman reported that she was floating above her own body and looking down on herself.

This makes some kind of neurological sense. The TPJ processes visual and touch signals, balance and spatial information from the inner ear, and the proprioceptive sensations from joints, tendons and muscles that tell us where our body parts are in relation to one another. Its job is to meld these together to create a feeling of embodiment: a sense of where your body is, and where it ends and the rest of the world begins. Blanke and colleagues hypothesized that out-of-body experiences arise when, for whatever reason, the TPJ fails to do this properly.

The shifting of the self

More evidence later emerged that a malfunctioning TPJ was at the heart of the out-of-body experience. In 2007, for example, Dirk De Ridder of University Hospital Antwerp in Belgium was trying to help a 63-year-old man with intractable tinnitus. In a last-ditch attempt to silence the ringing in his ears, Ridder's team implanted electrodes near the patient's TPJ. It did not cure his tinnitus, but it did lead to him experiencing something close to an out-of-body experience: he would feel his self shift about 50 centimetres behind and to the left of his body. The feeling would last more than 15 seconds, long enough to carry

FIGURE 7.2   Put yourself in his place: creating an
'own-body transformation'

out PET scans of his brain. Sure enough, the team found that
the TPJ was activated during the experiences.

Insights from neurological disorders or brain surgery can
only take you so far, however, not least because cases are rare.
Larger-scale studies are required, and to achieve this Blanke and
others have used a technique called 'own-body transformation
tasks' to force the brain to do things that it seemingly does during an out-of-body experience. In these experiments, subjects
are shown a sequence of brief glimpses of cartoon figures wearing a glove on one hand. Some of the figures face the subject,
others have their back turned (*see* diagram above).

The task is to imagine yourself in the position of the cartoon
figure in order to work out which hand the glove is on. To do
this, you may have to mentally rotate your own body as one
image succeeds another. As volunteers performed these tasks,
the researchers mapped their brain activity with an EEG and
found that the TPJ was activated when the volunteers imagined
themselves in a position different from their actual orienta-
tion – an out-of-body position.

The team also zapped the TPJ with **transcranial magnetic
stimulation**, a non-invasive technique that can temporarily dis-
able parts of the brain. With a disrupted TPJ, volunteers took
significantly longer to do the own-body transformation task.

Other brain regions have been implicated, too, including ones close to the TPJ. The emerging consensus is that when these regions are working well, we feel at one with our body. But disrupt them, and our sense of embodiment can float away.

This does not, however, explain the most striking feature of out-of-body experiences: why most people, from their out-of-body locations, visualize not only their bodies but things around them, such as other people. Where does this information come from?

Sleep paralysis

One line of evidence comes from the condition known as **sleep paralysis**, in which healthy people find their body immobilized as in sleep despite being conscious. In a survey of nearly 12,000 people who had experienced sleep paralysis, Allan Cheyne of the University of Waterloo in Ontario, Canada, found that many reported sensations similar to out-of-body experiences. These included floating out of their body and turning back to look at it.

Cheyne suggests that this might be the result of conflicts of information in the brain. During sleep paralysis, it is possible to enter an REM-like state in which you dream of moving or flying. Under these circumstances you are conscious of a sensation of movement, yet your brain is aware that your body cannot move. In an attempt to resolve this sensory conflict, the brain cuts the sense of self loose. Perhaps similar sensory conflicts cause classic out-of-body experiences.

Brugger, meanwhile, has a suggestion for how someone might see things even though their eyes are shut, based on one of his patients who reported an out-of-body experience. According to this patient's father, who was sitting by the bedside, he had his eyes closed. Yet he later reported seeing, from a perspective above his bed, his father going to the bathroom, returning with a wet towel and towelling his forehead.

The patient presumably heard his father walk to the bathroom and run some water, and must have felt the wet towel on his head. Brugger speculates that his brain converted those stimuli into a visual image, not unlike what happens in synaesthesia. This still does not, however, explain the external vantage point.

Thomas Metzinger of the Johannes Gutenberg University of Mainz, Germany, has a suggestion. Imagine an episode from a recent holiday. Do you visualize it from a first-person perspective, or from a third-person perspective with yourself in the scene? Surprisingly, most of us do the latter. If the brain is reconstructing information from memory, then it makes sense that it is from the point of view of the outside.

Whatever the mechanism, the study of out-of-body experiences promises to help answer a profound question in neuroscience and philosophy: how does self-consciousness emerge? It's abundantly clear to us that we have a sense of self that resides, most of the time, in our bodies. Yet it is also clear from out-of-body experiences that the sense of self can seemingly detach from your physical body. So how are the self and the body related?

To address that question, Metzinger teamed up with Blanke and his colleagues in an experiment that induces an out-of-body experience in healthy volunteers. They filmed each volunteer from behind and projected the image into a head-mounted display worn by the volunteer so that they see an image of themselves standing about 2 metres in front. The experimenters then stroke the volunteer's back – which the volunteers see being done to their virtual self. This creates sensory conflict, and many reported feeling their sense of self migrating out of their physical bodies and towards the virtual one.

To Metzinger, these experiments demonstrate that self-consciousness begins with the feeling of owning a body, but there is more to self-consciousness than the mere feelings of embodiment. Selfhood most likely has many components, Metzinger believes, and we are only beginning to piece them all together.

Epileptic bliss

It was one of the most profound experiences of Fyodor Dostoevsky's life. 'A happiness unthinkable in the normal state and unimaginable for anyone who hasn't experienced it... I am then in perfect harmony with myself and the entire universe,' the novelist told his friend, Russian philosopher Nikolai Strakhov. What lay behind such feelings? The description might suggest a religious awakening – but Dostoevsky was instead describing the moments before a full-blown epileptic seizure.

Those sensations seem to have informed the character of Prince Myshkin in Dostoevsky's novel, *The Idiot.* 'I would give my whole life for this one instant,' the prince says of the brief moment at the start of his epileptic fit – a moment 'overflowing with unbounded joy and rapture, ecstatic devotion, and completest life.'

In recent years, scientists have taken a renewed interest in reports of epileptic bliss as a window on self-awareness. They also want to know whether there might be other ways in which we could all be transported to similar states of being.

Epileptic seizures are broadly divided into two groups: generalized and focal. In generalized seizures, electrical discharges overwhelm the outer layer of the brain, the cortex, and often lead to loss of consciousness. Ecstatic seizures seem to be of the second kind. In focal or partial epilepsy the electrical storm is confined to a small region of the brain and the person usually remains conscious. This type of seizure can turn into a generalized one if the errant electrical signals spread.

Fabienne Picard, a neurologist at the University Hospital in Geneva, Switzerland, has interviewed many patients with ecstatic seizures and has identified three broad categories of effects.

The first was heightened self-awareness. For example, a 53-year-old female teacher told Picard: 'During the seizure it is as if I were very, very conscious, more aware, and the sensations, everything seems bigger, overwhelming me.' The second was a sense of physical well-being. A 37-year-old man described it as 'a sensation of velvet, as if I were sheltered from anything negative'. The third was intense positive emotions, best articulated by a 64-year-old woman: 'The immense joy that fills me is above physical sensations. It is a feeling of total presence, an absolute integration of myself, a feeling of unbelievable harmony of my whole body and myself with life, with the world, with the "All",' she said.

As Picard began looking for the neurological origin of the disorder, such descriptions pointed her towards the **insula** – a region of the cortex that is of growing interest to scientists studying consciousness. It is buried inside the fissure dividing the frontal and parietal lobes from the temporal lobe, and its main function seems to be to integrate 'interoceptive' signals from inside the body, such as the heartbeat, with 'exteroceptive' signals such as the sensation of touch.

There is also evidence that the processing of these signals gets progressively more sophisticated looking from the back of the insula to the front. The portion of the insula closest to the back of the head deals with objective properties, such as body temperature, and the front portion, or **anterior** insula, produces subjective feelings of body states and emotions, both good and bad. In other words, the anterior insula is responsible for how we feel about our body and ourselves, helping to create a conscious feeling of 'being'. This led Bud Craig at the Barrow Neurological Institute in Phoenix, Arizona, to argue that this part of the brain is the key to 'the ultimate representation of all of one's feelings – that is, the sentient self'.

The root of ecstatic epilepsy

Investigating how abnormal activity in the anterior insula leads to disorders like ecstatic epilepsy might also help scientists establish how this region creates our normal experience of self-awareness. Picard's patients reported feelings of certainty – the sense that all is right with the world – which would seem to fit with a theory that the anterior insula is involved in predicting the way the body is going to feel in the next instant. Those predictions are then compared with actual sensations, generating a 'prediction error' signal that might help to determine how we react to a changing environment. If the prediction error is small, we feel good, if it is large we feel anxious. It is possible that the electrical storm in the anterior insula may be disrupting the comparator mechanism, causing there to be no prediction error. As a result, the person is left feeling as if nothing is wrong with the world, that everything makes sense.

Besides the sense of expanded awareness and certainty, people like Dostoevsky have also recorded the strange sense that time is slowing down during their seizures. This might reflect the way the insula samples our senses. Bud Craig argues that the anterior insula usually combines interoceptive, exteroceptive and emotional states to create a discrete 'global emotional moment' every 125 milliseconds or so – dividing our feelings into separate frames, like a film reel. He posits that a hyperactive anterior insula may generate these global emotional moments faster and faster, leading to a sense that time is slowing.

We could also gain insights into the insula's role by other means. Craig and Picard think that feelings evoked by drugs like amphetamine, ecstasy and cocaine may share many similarities with ecstatic epilepsy. These chemicals usually trigger a flush of neurotransmitters through the brain, and there is evidence that, following drug use, levels of **dopamine** in the anterior insula are unusually high relative to other regions. The neurotransmitter **serotonin** may be similarly implicated in the case of *ayahuasca*, a psychedelic brew long associated with shamanistic rituals in the Amazon. Again, nuclear imaging results show increased blood flow in the anterior insula about 100 minutes after consumption.

Fortunately, there may be safer ways to come close to the same feelings. Meditators often experience the time-slowing, heightened self-awareness and feelings of profound well-being that come with Dostoevsky syndrome. In 2007, Richard Davidson of the University of Wisconsin–Madison and colleagues studied 15 expert and 15 novice meditators. They found that the deeper the meditative state, the greater the activity in the anterior insula.

If that does reflect the same 'unbounded joy and rapture' that Dostoevsky's Prince Myshkin reported, it certainly doesn't come easily: the experienced meditators had logged more than 10,000 hours of practice to see these effects. You may not need to give your 'whole life for this one instant', as Prince Myshkin put it, but it may not be far off.

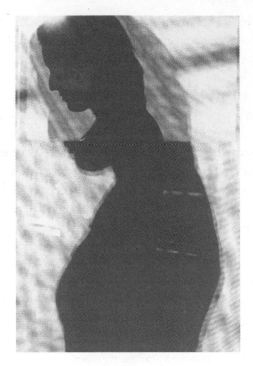

FIGURE 7.3 Some altered states of consciousness involve a
shifting of the self

Orgasm: a holiday from the mind

At the moment of orgasm, consciousness goes out of the window. But what is happening in our brains and bodies to make that happen? Barry Komisaruk at Rutgers University in Newark, New Jersey, and colleagues are trying to find out, teasing apart the mechanisms underlying sexual arousal. In doing so, not only have they discovered that there is more than one route to orgasm, but they may also have revealed a novel type of consciousness – an understanding of which could lead to new treatments for pain.

Komisaruk is interested in the time course of orgasm, and particularly when an area of the brain called the **prefrontal cortex** (PFC) becomes active. The PFC is situated at the front of the brain and is involved in aspects of consciousness, such as self-evaluation and considering something from another person's perspective.

Komisaruk's team recently found heightened activation in the PFC during female climax – something not seen in previous studies of the orgasm. Surprisingly, this was also the case in individuals who can achieve orgasm by thought alone. With fantasy and self-referential imagery often reported as being part of the sexual experience, Komisaruk and colleagues wondered if the PFC might be playing a key role in creating a physiological response from imagination alone.

In experiments, he has shown that the PFC is particularly active when someone imagines being sexually touched, rather than when they actually are. He suggests this heightened activation may reflect imagination or fantasy, or perhaps some cognitive process that helps manage so called 'top-down' control – the direct regulation by the brain of physiological functions – of our own pleasure.

However, when Janniko Georgiadis at the University of Groningen in the Netherlands, and colleagues, performed similar experiments they found that the same brain region 'switched off' during orgasm. Specifically, they saw significant deactivation in an area of the PFC called the **left orbitofrontal cortex** (OFC).

Georgiadis argues that the OFC may be the basis of sexual control – and perhaps only by letting go, so to speak, can orgasm be achieved. He suggests this deactivation may be the most telling example of an 'altered state of consciousness'; a loss of control that has not been seen, as yet, during any other type of activity.

There may be a simple explanation for the discrepancies between Georgiadis's and Komisaruk's work – they may represent two different paths to orgasm, activated by different methods of induction. While participants in Komisaruk's studies masturbated themselves to orgasm, those in Georgiadis's were stimulated by their partners. Perhaps having a partner makes it easier to let go of that control and achieve orgasm. Alternatively, having a partner may make top-down control of sensation and pleasure less necessary to climax.

Komisaruk agrees. He hopes to one day use **neurofeedback** to allow women who do not experience orgasm to view their brain activity in real time during genital stimulation. The hope is that this feedback may help them to manipulate their brain activity to bring it closer to that of an orgasmic pattern of activity. He also believes that further study of the orgasm – and the PFC's role – will offer much-needed insight into how we might use thought alone to control other physical sensations, such as pain. It may be that, wonderful as conscious control is, learning to leave it behind now and then may cure a multitude of ills.

How LSD expands your mind

Three-quarters of a century after chemist Albert Hofmann accidentally ingested LSD and experienced its mind-expanding effects, brain imaging has given researchers their first glimpse of how it causes its profound effects on consciousness.

One of the most notable aspects of the psychedelic experience is a phenomenon known as the 'dissolution of the ego', in which users feel somehow detached from themselves. Studying how the normally stable sense of self gets disrupted can tell us how neural mechanisms create this integral part of the human experience.

Robin Carhart-Harris of Imperial College London gave 20 volunteers infusions on two days, once containing 75 micrograms of LSD, the other a placebo. Then volunteers lay in a scanner and had their brains imaged with three different techniques, which together built up a comprehensive picture of neural activity, both with the drug and without.

Drug-free lala land

Fancy experiencing an altered state of consciousness without resorting to hallucinogenic drugs? No problem. We slip into such a state every night when we sleep, although it's hard to appreciate it at the time because of the inconvenient fact that we are unconscious.

By paying closer attention to the half-way stage as you drop off, known as **hypnagogia**, when hallucinations are surprisingly common, it might be possible to access this out-of-body state.

We don't know what causes hypnagogia but one theory is that some parts of the brain are falling asleep ahead of the rest. This nightly altered state can inspire creativity. Chemist Friedrich August Kekulé had his insight about the ring structure of benzene while half-asleep. Then there was the surrealist Salvador Dalí, who tuned in to his creativity by letting himself drop off while suspending a spoon over a metal plate. As he fell asleep, the spoon would crash down, jerking him awake while the dream images were still fresh in his mind.

Hypnagogia is not all fun though. It can sometimes trigger one terrifying kind of sleep paralysis, when the nerve inhibition that normally accompanies our dreams kicks in before someone is fully asleep. This state can often be accompanied by threatening auditory and visual hallucinations and is thought to account for reports of alien abduction.

MRI scans showed that LSD caused brain activity to become less coordinated in regions that make up what is called the **default mode network**. The size of the effect was correlated with participants' ratings of their own ego dissolution, suggesting that this network underlies a stable sense of self.

Another imaging type, **magnetoencephalography** (MEG), showed that the rhythm of alpha brainwaves weakened under LSD, an effect that was also correlated with ego dissolution. Alpha rhythms are stronger in humans than other animals, and Carhart-Harris thinks it could be a signature of high-level human consciousness.

But LSD also made the brain more unified in its activity, and there was more communication between regions that normally work separately. This suggests that the brain is functioning more simply than usual.

The results also go some way to explaining how LSD causes dreamlike visual hallucinations. Although the primary visual cortex usually communicates mainly with other parts of the vision system, many other brain areas contributed to the processing of images in volunteers who received LSD.

There was intense research into LSD in the 1950s and 1960s, and the drug showed great promise in treating mood disorders, addictions and other conditions. When it was banned by an international treaty, most scientific work ground to a halt even though it was still technically allowed. David Nutt, the senior author of the study, says he hopes the study will be transformative and inspire others to follow them in the search for consciousness via altered states of mind.

Psychedelic drugs push brain into a state never needed before

Measuring neuron activity has revealed that psychedelic drugs really do alter the state of the brain, creating a different kind of consciousness.

Anil Seth, at the University of Sussex, UK, discovered this by re-analysing data previously collected by researchers at Imperial College London. Robin Carhart-Harris and his colleagues had monitored brain activity in 19 volunteers who had taken ketamine, 15 who had had LSD, and 14 who were under the influence of **psilocybin**, a hallucinogenic compound in magic mushrooms. Carhart-Harris's team used sets of sensors attached to the skull to measure the magnetic fields produced by these volunteers' neurons, and compared these to when each person took a placebo.

Previous work had shown that people in a state of wakefulness have more diverse patterns of brain activity than people who are asleep. Seth's team has found that people who have taken psychedelic drugs show even more diversity – the highest level ever measured.

These patterns of very high diversity coincided with the volunteers reporting 'ego-dissolution' – a feeling that the boundaries between oneself and the world have been blurred. The degree of diversity was also linked to more vivid experiences.

There's mounting evidence that psychedelic drugs may help people with depression in ways that other treatments can't. Some benefits have already been seen with LSD, ketamine, psilocybin, and ayahuasca, a potion used in South America during religious rites.

Interview: Why we should destigmatize hallucinations

Oliver Sacks, the celebrated neurologist and life-long champion of people who experienced altered states of consciousness, died in 2015. A few years prior to his death he spoke to *New Scientist* about why we should embrace our altered states, rather than be afraid of them.

What interests you about hallucinations?

I've been fascinated with them for a long time. There's such a vast variety, and there are so many causes, so much misunderstanding – and sometimes so much stigma attached – that I thought it would be good to bring things together. An additional reason has been the beautiful neuroimaging in the past ten years or so, which confirms that at least simple hallucinations tend to arise in sensory areas which normally serve perception.

You mentioned stigma. Do most people associate hallucinations with mental illness?

I think there's a common view, often shared by doctors, that hallucinations denote madness – especially if there's any hearing of voices. I hope I can defuse or destigmatize this a bit. This can be felt very much by patients. There was a remarkable study of elderly people with impaired vision, and it turned out that many had elaborate hallucinations, but very few acknowledged anything until they found a doctor whom they trusted.

What is the difference between hallucination and imagination?

I think you recognize that what you imagine is your own, whereas with hallucinations there is no sense of you having produced them. One feels, 'What's that? Where did it come from?'

I saw this very clearly many years ago in an old lady who started to hear Irish songs in the middle of the night. She thought a radio had been left on but couldn't find the radio. She then thought that a tooth filling was somehow acting as a transistor. Finally, when certain tunes kept repeating themselves, all tunes that she knew, she wondered if it was a sort of radio inside her head, a mechanism not under her control, and apparently not related to what she was thinking or feeling or doing. That way of putting things is very common in people with musical hallucinations.

In your book, *Hallucinations*, you share experiences of your 'lost years' in California in the early 60s, when you tried lots of drugs. Why write about this now?

The primary reason is that what happens with me is a potential source of information. I will, as it were, use my own case history as I will use other people's. But perhaps again the fact that these were encapsulated in a time period more than 40 years ago made me feel easier about describing them.

You experimented with LSD and other hallucinogenics. Have those experiences informed your work as a neurologist?

I think it made me more open to some of my patients' experiences. For example, there is something which I think of as stroboscopic vision, or cinematic vision, where, instead of seeing a scene continuously, you see a series of stills. I've had that myself on LSD, I've had it in migraine, and my patients taking L-dopa sometimes describe it, too. So rather than saying nonsense, or closing my ears, I am open to these descriptions. Whether these psychedelic drugs made much difference to me otherwise, I don't know. I'm glad I had the experience. It taught me what the mind is capable of.

One time you had a conversation with a spider...

With the spider, I should have known that it's impossible. That's one of the few times when I was completely taken in. The business of believing and being converted by hallucinations worries me. For example, a neurosurgeon who had a so-called near-death experience published a book saying that and is convinced that he saw Heaven. I want to say, strongly, hallucinations aren't evidence of anything, let alone Heaven.

You highlight a tendency for hallucinations, particularly those caused by epileptic seizures, to feel like religious experiences. Why is that?

Hallucinations can be very powerful and very persuasive. I think one may have to fight to deny them weight. There was one case history which I should have put in the book. A young woman, a physician, had some of these seemingly revelatory seizures, but she argued with God. God said: 'Don't you believe your senses?' She said: 'Not when I'm in a seizure.'

Do you worry that sharing your patients' stories somehow exploits them?

I'm on this delicate boundary, and have been for 50 years or so. At one time I was my own prime accuser. Whenever I saw the word portrayal, I would misread it as betrayal. First, in addition to any formal consent, I want to be reassured from what I know of a patient that they won't be upset by anything.

Do you hope that sharing these stories changes people's perceptions?

I feel that if I describe things respectfully, tenderly and truly, then this is an important thing to do. It's not voyeurism, it's not exploitation, but an essential form of knowledge. I think the detailed case history has no equal in conveying understanding, not only of what a condition is like, but of the ways in which a person may respond to a condition.

I remember when an opera was made from my book *The Man Who Mistook his Wife for a Hat*, I said to the librettist, you must go and see Mrs P — the woman who was mistaken for a hat — and see how she would feel about this. I watched her watch the opera, wondering fearfully what she might be thinking. But she came up to me and the librettist, and said, you have done honour to my husband. I hope in some sense I can do honour to the patients.

8
Losing it

From sleep to anaesthesia, we all lose consciousness from time to time. Where do we go, exactly? And can this shed light on the phenomena of self-awareness?

The mystery of anaesthesia

The development of general anaesthesia has transformed surgery from an horrific ordeal into a gentle slumber. It is one of the commonest medical procedures in the world, yet we still don't know how the drugs work. Perhaps this isn't surprising: since we still don't understand consciousness, how can we comprehend its disappearance?

That is starting to change, however, with the development of new techniques for imaging the brain or recording its electrical activity during anaesthesia.

Altered consciousness doesn't only happen under a general anaesthetic of course – it occurs whenever we drop off to sleep, or if we are unlucky enough to be whacked on the head. But anaesthetics do allow neuroscientists to manipulate our consciousness safely, reversibly and with exquisite precision.

It was a Japanese surgeon who performed the first-known surgery under anaesthetic, in 1804, using a mixture of potent herbs. In the west, the first operation under general anaesthetic took place at Massachusetts General Hospital in 1846. A flask of sulphuric ether was held close to the patient's face until he fell unconscious.

Since then a slew of chemicals have been co-opted to serve as anaesthetics, some inhaled, like ether, and some injected. The people who gained expertise in administering these agents developed into their own medical speciality. Although long overshadowed by the surgeons who patch you up, the humble 'gas man' does just as important a job, holding you in the twilight between life and death.

Dimmer switch for consciousness

Consciousness is often thought of as an all-or-nothing quality – either you're awake or you're not – but in reality there are different levels of anaesthesia which act like a dimmer switch on our consciousness (*see* illustration below).

A typical subject first experiences a state similar to drunkenness, which they may or may not be able to recall later, before falling unconscious, which is usually defined as failing to move in response to commands. As they progress deeper into the twilight zone, they now fail to respond to even the penetration of a scalpel – which is the point of the exercise, after all – and at the deepest levels may need artificial help with breathing.

These days anaesthesia is usually started off with injection of a drug called propofol, which gives a rapid and smooth transition to unconsciousness. (This is also what Michael Jackson was allegedly using as a sleeping aid, with such unfortunate consequences.) Unless the operation is only meant to take a few minutes, an inhaled anaesthetic, such as isoflurane, is then usually added to give better minute-by-minute control of the depth of anaesthesia.

How anaesthetics work

So what do we know about how anaesthetics work? Since they were first discovered, one of the big mysteries has been how the members of such a diverse group of chemicals can all result in the loss of consciousness. Other drugs work by binding to receptor molecules in the body, usually proteins, in a way that relies on the drug and receptor fitting snugly together like a key in a lock. Yet the long list of anaesthetic agents ranges from large complex molecules such as barbiturates or steroids, to the inert gas xenon, which exists as mere atoms. How could they all fit the same lock?

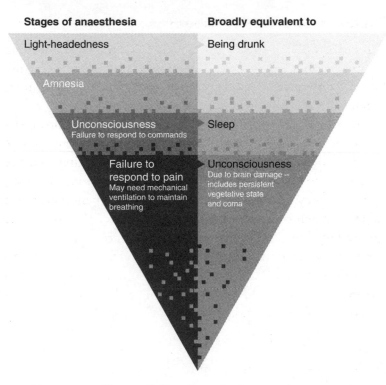

Stages of anaesthesia

Light-headedness

Amnesia

Unconsciousness
Failure to respond to commands

Failure to respond to pain
May need mechanical ventilation to maintain breathing

Broadly equivalent to

Being drunk

Sleep

Unconsciousness
Due to brain damage – includes persistent vegetative state and coma

FIGURE 8.1 You are feeling sleepy: Losing consciousness under anaesthesia is not so much flipping a light switch as turning down a dimmer switch

For a long time there was great interest in the fact that the potency of anaesthetics correlates strikingly with how well they dissolve in olive oil. The popular 'lipid theory' said that instead of binding to specific protein receptors, the anaesthetic physically disrupted the fatty membranes of nerve cells, causing them to malfunction.

In the 1980s, though, test-tube experiments showed that anaesthetics could bind to proteins in the absence of cell membranes. Since then, protein receptors have been found for many anaesthetics. Propofol, for instance, binds to receptors on nerve cells that normally respond to a chemical messenger called GABA. Presumably the solubility of anaesthetics in oil affects how easily they reach the receptors bound in the fatty membrane.

But that solves only a small part of the mystery. We still don't know how this binding affects nerve cells, and which neural networks they feed into.

Many anaesthetics are thought to work by making it harder for neurons to fire, but this can have different effects on brain function, depending on which neurons are being blocked. So brain-imaging techniques such as functional MRI scanning, which tracks changes in blood flow to different areas of the brain, are being used to see which regions of the brain are affected by anaesthetics. Such studies have been successful in revealing several areas that are deactivated by most anaesthetics. Unfortunately, so many regions have been implicated it is hard to know which, if any, are the root cause of loss of consciousness.

Taking the global workspace theory of consciousness (*see* Chapter 2) this is perhaps not surprising. This states that incoming sensory information is first processed unconsciously in separate brain regions. We only become conscious of the experience if these signals are broadcast to a network of neurons spread through the brain, which then start firing in synchrony.

Fading awareness

The idea has recently gained support from recordings of the brain's electrical activity using electroencephalograph (EEG) sensors on the scalp as people are given anaesthesia. This has shown that as consciousness fades there is a loss of synchrony between different areas of the cortex – the outermost layer of the brain important in attention, awareness, thought and memory.

This process has also been visualized using fMRI scans. Steven Laureys, who leads the Coma Science Group at the University of Liège in Wallonia, Belgium, looked at what happens during propofol anaesthesia when patients descend from wakefulness, through mild sedation, to the point at which they fail to respond to commands. He found that while small 'islands' of the cortex lit up in response to external stimuli when people were unconscious, there was no spread of activity to other areas, as there was during wakefulness or mild sedation.

A team led by Andreas Engel at the University Medical Center in Hamburg, Germany, has been investigating this process in still more detail by watching the transition to unconsciousness in slow motion. Normally it takes about ten seconds to fall asleep after a propofol injection. Engel has slowed it down to many minutes by starting with just a small dose, then increasing it in seven stages. At each stage he gives a mild electric shock to the volunteer's wrist and takes EEG readings.

We know that upon entering the brain, sensory stimuli first activate a region called the primary sensory cortex, which runs like a headband from ear to ear. Then further networks are activated, including frontal regions involved in controlling behaviour, and temporal regions towards the base of the brain that are important for memory storage. Engel found that at the deepest levels of anaesthesia, the primary sensory cortex was the only region to respond to the electric shock. The message seemingly never got as far as the global workspace.

What could be causing the blockage? Engel has unpublished EEG data suggesting that propofol interferes with communication between the primary sensory cortex and other brain regions by causing abnormally strong synchrony between them which leaves no room for more subtle messages. Something similar happens in epilepsy seizures that cause a loss of consciousness.

Feedback loops

Experiments have also shown that anaesthetic disrupts the two-way communication needed for information to be integrated in the brain. George Mashour, an anaesthetist at the University of Michigan in Ann Arbor, and his group published EEG work showing that both propofol and the inhaled anaesthetic sevoflurane inhibited the transmission of feedback signals from the frontal cortex in anaesthetized surgical patients. The backwards signals recovered at the same time as consciousness returned. This is support for the idea that consciousness depends heavily on the activity of feedback loops between different areas of the brain.

Similar findings are coming in from studies of people in a coma or persistent vegetative state (PVS), who may open their eyes in a sleep-wake cycle, although remain unresponsive. Laureys, for example, has seen a similar breakdown in communication between different cortical areas in people in a coma.

Adrian Owen of the University of Western Ontario in London, Canada, hopes that studies of anaesthesia will shed light on disorders of consciousness such as coma. Owen and others have previously shown that people in a PVS respond to speech with electrical activity in their brain. More recently he did the same experiment in people progressively anaesthetized with propofol. Even when heavily sedated, their brains responded to speech. But closer inspection revealed that those parts of the brain that decode the meaning of speech had indeed switched off, prompting a rethink of what was happening in people with PVS.

Just how anaesthesia allows us to take a holiday from awareness for a short and controlled period of time remains to be fully explained. Without really understanding how, anaesthetists guide hundreds of millions of people a year as close to the brink of nothingness as it is possible to go without dying. Then they bring them safely back home again.

Are you awake? Exploring the mind's twilight zone

In 2009, a puzzling report appeared in the journal *Sleep Medicine*. It described two Italian people who never truly slept. They might lie down and close their eyes, but read-outs of brain activity showed none of the normal patterns associated with sleep. Their behaviour was pretty odd, too. Though largely unaware of their surroundings during these rest periods, they would walk around, yell, tremble violently and their hearts would race. The remainder of the time they were conscious and aware but prone to powerful, dream-like hallucinations.

Both had been diagnosed with a neurodegenerative disorder called **multiple system atrophy**. According to the report's authors − Roberto Vetrugno and colleagues from the University of Bologna, Italy − the disease had damaged the pair's brains to such an extent that they had entered **status dissociatus**, a kind of twilight zone in which the boundaries between sleep and wakefulness completely break down.

That this can happen contradicts the way we usually think about sleep, but it came as no surprise to Mark Mahowald, medical director of the Minnesota Regional Sleep Disorders Center in Minneapolis, who has long contested the dogma that sleep and wakefulness are discrete and distinct states. The blurring of sleep and wakefulness is very clear in status dissociatus, but he believes it can happen to us all. If he is right, we will have to rethink our understanding of what sleep is and what it is for. Maybe wakefulness is not the all-or-nothing phenomenon we thought it was.

Received wisdom has it that at any given time, healthy people are in one of three states of vigilance: awake, in rapid eye movement (REM) sleep or in non-REM (NREM) sleep. Each state is distinct and can be recognized by a characteristic

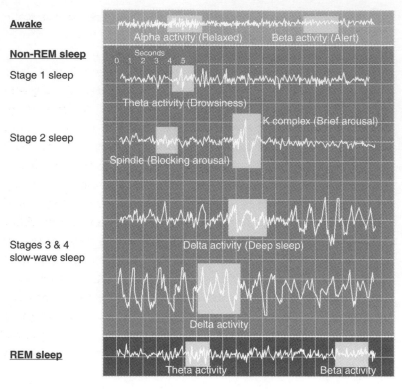

FIGURE 8.2 States of vigilance: The traditional way of distinguishing wakefulness from non-REM and REM sleep is to look for characteristic patterns of brain activity using an EEG, which measures electrical activity in the outer few millimetres of the cortex

pattern of brain activity, as measured by an electroencephalogram (EEG) (*see* Figure 8.2).

Wakefulness is easy to detect. Apart from the fact that a person's eyes are open and they are responsive, their EEG shows a pattern of high-frequency, low-amplitude waves. NREM sleep is divided into four stages, each of which has its own distinctive EEG pattern. REM is trickier to spot because in EEG terms it closely resembles stage 1 NREM sleep. So to be sure it really is REM, researchers also look for the tell-tale rapid eye movements and a slackening in the muscles of the chin and jaw.

Mahowald is not the only person to have questioned these neat distinctions. David Dinges, a psychiatrist at the University of Pennsylvania, Philadelphia, has probably deprived more people of sleep in the name of science than anyone else. In one such study in the late 1980s, Dinges and his team revealed how easily the different states of vigilance can become intermingled. When volunteers were subjected to tests of working memory in which they had to subtract numbers, they could do an average of 90 sums in 3 minutes with few errors. After 52 hours deprived of sleep, their performance fell to around 70 subtractions, with not many more errors. However, after they had slept for 2 hours the change was dramatic. People may have rated themselves as alert, but they couldn't manage even the simplest subtractions. People even seemed to be dreaming as they attempted the task. One subject mused: 'What if people ran faster than normal people run home,' in the middle of a string of incorrect responses.

Sleep inertia

Known as **sleep inertia**, a less extreme version of such disorientation is now generally recognized as the cause of the grogginess some people get after their alarm clock goes off. It is as if they are socially awake but functionally asleep; as if the brain circuits underlying responsiveness are up and running, but those mediating working memory are still offline.

Many sleep disorders may also result from a blurring of the lines. One is **REM behavioural disorder** (RBD), in which people in REM sleep act out their dreams because the temporary paralysis, or **cataplexy**, that normally accompanies this state is replaced by the full mobility of wakefulness. In sleep paralysis the opposite is true. Here, cataplexy intrudes into wakefulness, and a person wakes to find himself or herself unable to move. It is estimated that up to 40 per cent of people have experienced this disturbing phenomenon. Also

FIGURE 8.3 When we fall asleep, we enter the mind's twilight zone

surprisingly common are **hypnagogic hallucinations** – sensory illusions that occur on the cusp of sleep when the dreaming component of REM intrudes into wakefulness. Other sleep disorders that come under this umbrella include **sleepwalking**, **night terrors** and **narcolepsy**, which is an inherent instability in vigilance state boundaries characterized by rapid cycling between states and the tendency to fall asleep midsentence. Perhaps surprisingly, it might also account for neardeath experiences and alien abductions. It is no coincidence, says Mahowald, that alien abductions almost always occur in the transition from wakefulness to sleep.

Microsleeps

The boundaries between sleep and wakefulness are particularly blurred when we are sleep-deprived. Around a decade ago, Dinges realized that although his sleep-deprived volunteers seemed to be awake, they were in fact experiencing momentary lapses, or **microsleeps**. Since then, he has discovered that these fleeting naps last between half a second and 2 seconds, and become increasingly frequent the longer we are deprived of sleep, until finally we cannot recover and nod off. Dinges sees them as the outward sign of a tug-of-war between neural systems that are trying to initiate sleep, and others that are trying to maintain wakefulness.

This chimes with the ideas of James Krueger of Washington State University in Pullman, who has argued that the individual processing units in the brain – known as **cortical columns** – fall asleep independently when they become tired. In his view, shifts between wakefulness and sleep occur when enough columns are in one state or the other. Krueger believes this mosaic pattern of sleep explains sleep inertia and sleepwalking.

Some people are more prone to microsleeps than others. In a 2007 study, Dinges and his colleagues showed that there are enormous differences in people's ability to resist the lure of sleep when tired. Among a group of healthy, non-sleep-deprived adults, the differences in alertness are small. Keep them awake for long periods and those differences grow.

Brain imaging has recently revealed a mental back-up system in people who remain alert when sleep-deprived. While other people have reduced brain activity when tired, sleep-resistant individuals manage to maintain their brain activity levels. More interestingly, they also recruit new areas to help compensate for having been awake for so long. These people were selected for the study because they had a gene variant found in around 40 per cent of people that is thought to be associated with resistance to sleep deprivation. It seems likely that such people are also less prone to state dissociation, although this has not yet been tested. Most of us, however, don't skip sleep without losing our grip on consciousness.

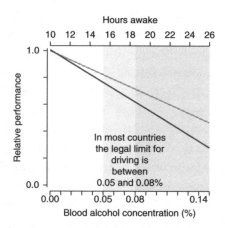

FIGURE 8.4 Drunk on fatigue. After about 17 hours awake, your cognitive and motor skills resemble those of someone who is intoxicated

Another group that appears to be more vigilant than most are people who suffer from **insomnia**. There is evidence to suggest they are in a constant state of hyperarousal, with relatively high metabolic rates and high levels of the stress hormone cortisol.

As the blur between sleep and wakefulness becomes more widely accepted, researchers are devising techniques for capturing the brain's fleeting lapses and vascillations. For example, neuroscientist Giulio Tononi, of the University of Wisconsin--Madison, is eavesdropping on sleeping brains using EEGs with 256 electrodes, rather than the more usual 32, to improve spatial resolution and help him catch the brain in the act of mosaic napping. 'Microsleeps are just the tip of the iceberg,' Tononi says. He is particularly concerned by the possibility that parts of our brain might be going offline without us even realizing it. Forgetfulness and daydreaming could be examples of this, but so could more bizarre and even criminal behaviours

Meanwhile, Pierre Maquet at the University of Liège in Belgium has started to use fMRI to chart the patterns of brain activity associated with different sleep states. His team is already finding that the distinctions between sleep and wakefulness look quite different when you compare patterns of activity across the whole brain, including deep structures, as opposed to using EEG, which measures activity in just the outer few millimetres of the cortex.

Lurking in the background is the hope that these approaches will shed light on the vexed issue of what sleep is for. A leading theory is that it is important for memory consolidation. Yet one of the puzzling aspects of the two Italians with status dissociatus is that, despite complete disruption of both their REM and NREM sleep, they showed no memory deficit. Does this add grist to the mill of those who believe that sleep has no other purpose than to save energy and keep us safe? Or does it mean, as Mahowald believes, that the two individuals were in fact experiencing some sort of mosaic sleep? Armed with a less black-and-white definition of sleep and wakefulness, and more sensitive tools for measuring them, that question might finally be resolved.

Dreaming: where do our minds go at night?

Some of our greatest mental adventures take place while we are unconscious: fast asleep and dreaming. Yet this internal fantasyland is a fickle beast. Anyone who has ever awoken feeling amazed by their night's dream only to forget its contents by the time they reach the shower will understand the difficulties of grasping such an ephemeral state of mind for long enough to understand it. Nevertheless, attempts to make sense of the dreams of large numbers of people are starting to throw some light on where we go at night and what it means for our minds in general.

Attempts to catalogue dream features involve asking participants to jot them down as soon as they wake up every morning or, better still, to sleep in a lab, where they are awoken and immediately questioned at intervals in the night. Such experiments have shown that our dreams tend to be short on sensory awareness. Most are silent movies – just half contain traces of sounds and taste, smell and touch appearing only very rarely.

Another approach is to look at the brain's activity during sleep for clues to the making of our dreams. Of particular interest is the idea that sleep helps to cement our memories for future recall. After first recording an event in the hippocampus – which can be thought of as the human memory's printing press – the brain then transfers its contents to the cortex, where it files the recollection for long-term storage.

Memories resurfacing

This has led some psychologists to suspect that certain elements of the memory may surface in our dreams as the different pieces of information are passed across the brain. Studying participants' diaries of real-life events and comparing them with their dream records, his team has found that memories enter our dreams in two separate stages. They first float into our consciousness on the night after the event itself, which might reflect the initial recording of the memory, and then they reappear between five and seven days later, which may be a sign of consolidation.

Even so, it is quite rare for a single event to appear in a dream in its entirety – instead, our memories emerge as small fragments that get incorporated into the story of the dream. The order in which the different elements appear might reflect the way a memory is broken down and then repackaged during consolidation.

Patrick McNamara, a neuroscientist at Northcentral University in Prescott Valley, Arizona, compared one individual's dream and real-life diaries over a two-month period. He found that a sense of place – a recognizable room, for instance – was the first fragment of a memory to burst onto the subject's dreamscape, followed by characters, actions and finally physical objects.

While it may cement a memory into our synapses during consolidation, the sleeping brain also forges links to other parts of your mental autobiography, allowing you to see associations between different events. This might dredge up old memories and plant them in our dreams, which in turn might explain why we often dream of people and places that we haven't seen or visited for months or even years. It could also lie behind those bizarre cases of mistaken identity while dreaming, when objects or people can appear to be one thing, but assume another shape or character.

Emotional undercurrents

Our dreams are more than a collection of characters and objects, of course. Like films or novels, they tell their stories in many different styles – from a trivial and disordered sequence to an intense poetic vision. Our emotional undercurrents seem to be the guiding force here. Ernest Hartmann, a psychiatrist at Tufts University in Medford, Massachusetts, has studied the dream diaries of people who have recently suffered a painful personal experience or grief. He found that they are more likely to have particularly vivid dreams that focus on a single central image, rather than a meandering narrative. These dreams are also more memorable than those from other, more placid times.

Hartmann suspects this might also reflect underlying memory processes – our emotions, after all, are known to guide which memories we store and later recall. Perhaps the intense images are an indication of what a difficult process it is integrating a traumatic event with the rest of our autobiography. The result may help us to come to terms with that event.

Despite these advances, many, many mysteries remain. Top of the list is the question of the purpose of our dreams: are they essential for preservation of our memories, for instance – or could we manage to store our life's events without them? That, we don't know. But understand their origins, says McNamara, and we would get a better grasp on consciousness in general.

Then there's the impact of our lifestyles on our night-time consciousness, with some research suggesting that TV and movies may have caused a major shift in the form and content of our dreams (*see* Monochrome or technicolor?). If a few hours of television a day can change the nature of our dreams, just imagine what our intense relationships with computers are doing. Eva Murzyn at the University of Derby, UK, for instance, has found that people who take part in the *World of*

Warcraft online role-playing game incorporate its user-interface into their midnight adventures.

Monochrome or technicolor?

Strong hints that technology drives our dreams emerged with puzzling reports in the 1950s that most people dream in black and white. Why? Curiously, this seemed to change over the following decade, and by the late 1960s the majority of people in the west seemed to dream in colour. What could cause the transformation? Eva Murzyn at the University of Derby, UK, puts it down to changes in broadcasting – movies and TV burst into colour at about the same time as a generation's dreams emerged from greyscale. Intriguingly, she has found that a difference still lingers to this day – those born before the advent of colour TV are still more likely to report dreaming in black and white than those born afterwards.

Inspired to look into it by her own son's gaming, Jayne Gackenbach at Grant MacEwan University in Edmonton, Canada, has found that players are beginning to report a greater sense of control over their dreams, with the feeling that they are active participants inside a virtual reality. She points out that gamers are more likely to try to fight back when they dream of being pursued by an enemy, for instance. Ironically, this interaction seems to make even nightmares less scary and more exciting.

If you're after a more peaceful night, you might want to take inspiration from Hervey de Saint-Denys, an early dream researcher in the 19th century who found that certain scents could direct his dreams. To prevent his own expectations from

clouding the results, he asked his servant to sprinkle a few drops
of perfume on his pillow on random nights as he slept. Sure
enough, he found that it led his dreams to events associated
with that particular scent. More generally, recent studies con-
firm that sweet smells can spark emotionally positive dreams.

Then again, you may prefer to let your subconscious direct
your nightly wanderings. As unsettling and upsetting as they
can sometimes be, it is their mystery that makes dreams
so enchanting.

Weird dream? Your brain won't even try to describe it

You open your front door to find your boss – who is also a
cat. The bizarre can seem completely normal when you're
dreaming, perhaps because parts of your brain give up trying to figure out what's going on.

Armando D'Agostino of the University of Milan in
Italy thinks that the strangeness of dreams resembles psychosis, because individuals are disconnected from reality
and have disrupted thought processes that lead to wrong
conclusions. Hoping to learn more about psychotic
thoughts, D'Agostino and his colleagues investigated how
our brains respond to the bizarre elements of dreams.

Because it is all but impossible to work out what a person is dreaming about while they're asleep, D'Agostino's
team asked 12 people to keep diaries in which they were
to write detailed accounts of seven dreams. When volunteers could remember one, they were also told to record
what they had done that day and come up with an unre-
lated fantasy story to accompany an image they had been
given.

Using a 'bizarreness' scoring system, the researchers found that dreams were significantly weirder than the waking fantasies the volunteers composed. A month later, the reports were read back to each of the dreamers while their brain activity was monitored with an fMRI scanner. Both dreams and fantasies seemed to selectively activate a set of structures in the right hemisphere of the brain associated with complex language processing, such as understanding multiple meanings of a word.

Curiously, the activity in this area appeared to decrease as the narrative became increasingly bizarre. It is almost as if the brain is giving up trying to make sense of the dream, says D'Agostino.

9
The unconscious mind

Humans are rather proud of the powers of our conscious mind. Which might make it slightly unsettling to know that much of our internal life happens outside of consciousness. Welcome to the realms of the unconscious mind.

Modern notions of powerful 'subconscious' were invented by Sigmund Freud as part of his theory of **psychoanalysis**. Freud famously considered it to be a cesspit of our most basic animalistic desires that was engaged in a constant tug-of-war with the more logical and detached conscious mind.

This is a view that modern neuroscientists definitely don't share, and these days the subconscious is on a firmer scientific footing – although many neurobiologists avoid the word 'subconscious', preferring 'non-conscious', 'pre-conscious' or 'unconscious' to describe thought processes that happen outside consciousness.

They do agree with Freud on one thing, however. Our brains have an uncanny knack for making sense of the world, with no need for conscious involvement. Far from being a kind of feral autopilot that needs to be controlled it is a purposeful, active and independent guide to behaviour.

So what goes on in the murky depths of our minds, and what kinds of things go on in there that we have no idea about?

FIGURE 9.1 There's more to cognition than meets the eye

Decision-making

Wouldn't it be great if you could leave difficult decisions to your subconscious, secure in the knowledge that it would do a better job than conscious deliberation? Ap Dijksterhuis of Radboud University Nijmegen in the Netherlands proposed this counter-intuitive idea in the early 2000s and it became instantly popular.

Dijksterhuis had found that volunteers asked to make a complex decision – such as choosing between different apartments based on a baffling array of specifications – made better choices after being distracted from the problem before deciding. He reasoned that this is because unconscious thought can move beyond the limited capacity of working memory, so it can process more information at once.

The idea has been influential, but it may be too good to be true. Many subsequent studies have failed to replicate Dijksterhuis's results. And a recent analysis concluded that there is little reason to think the unconscious is the best tool for making complex decisions. Still, Dijksterhuis remains confident that the effect is real and is an important part of our mental toolkit.

Others think the unconscious mind's way of processing information is more important for creativity than for decision-making. It brings together disparate information from all over the brain without interference from the brain's goal-directed frontal lobes. This allows it to generate novel ideas that burst through to consciousness in a moment of insight. John Kounios of Drexel University in Philadelphia believes an idea can only be truly creative if it appears in this way.

Some people seem to be better wired for this kind of thinking. Kounios has found that people who tend to solve problems in 'aha' moments of insight have different resting state brain activity – with less frontal control – than more logical thinkers.

While there is no known way to change your brain into a more creative one, Kounios suggests thinking about a problem until you get stuck, then taking a break and hoping that something useful bubbles up before your deadline.

Predicting the future

Every moment, the brain takes in far more information than it can process on the fly. In order to make sense of it all, the brain constantly makes predictions that it tests by comparing incoming data against stored information. All without us noticing a thing.

Simply imagining the future is enough to set the brain in motion. Imaging studies have shown that when people expect a sound or image to appear, the brain generates an anticipatory signal in the sensory cortices.

This ability to be one step ahead of the senses has an important role in helping us understand speech. The brain is continuously one step ahead in the conversation, predicting the sounds, words and meanings that are likely to come out of someone's mouth next.

Studies have also shown that the brain can use one sense to inform another. When you hear a recording of speech that is so degraded it is nearly unintelligible, the words sound clearer if you have previously read the same words in subtitles. This shows that the sensory parts of the brain are comparing the speech you've heard to the speech you predicted based on previous knowledge.

Not only do we make hypotheses about external information, our brains also make predictions on the basis of emotional signals coming from the body. Moshe Bar, a neuroscientist at Bar-Ilan University in Israel goes so far as to suggest that we only consciously recognize an object once our unconscious mind has calculated its importance based on what our senses and emotional reaction are saying. The conscious fear of a snake on a hiking trail comes *after* the brain has processed the shape and initiated jumping out of the way, for example.

Making predictions does have its downsides, however. Incorrect inferences, reinforced by repetition, can be hard to reverse, which is why when you learn the wrong lyrics to a song, it can be difficult to stop hearing them. **Stereotyping** is a more troublesome example of the same thing. While it can be useful to recognize that the dangers of things like snakes and fires are relatively constant, when it comes to human interactions, it can lead to negative biases and discrimination.

Some neuroscientists also believe that the hallucinations experienced in psychosis are the result of expectations gone awry. In one recent study, people who were more prone to psychotic experiences were better at seeing hidden shapes in images that had been digitally degraded. The researchers speculate that this could mean their brains jump to conclusions faster and rely less on evidence coming in from the senses.

Despite its flaws, prediction is hugely beneficial. Without it we'd have to learn every lesson as we did the first time: the hard way.

How to reach the unconscious

Any investigation of unconscious mind is hampered by one thing: by definition, unconscious thoughts are the ones that we are unaware of. No matter how hard they try, people can't tell you about something of which they are not aware, and there is no way to tell unconscious from conscious processing on a brain scan. Instead, neuroscientists and psychologists have had to develop ingenious – and somewhat sneaky – ways to access them.

One is to study people with brain damage who have conditions such as **blindsight** – where they are unable

---

to see visual stimuli on one side following an injury or stroke. While they may be unaware of seeing an action or an object, if forced to guess what was in that visual field they perform far better than would be expected by chance. This suggests that while they may not consciously see a stimulus, they are able to unconsciously process what they have seen and respond appropriately.

Another approach, developed by Stanislas Dehaene, director of the Cognitive Neuroimaging Unit at INSERM, France, is called **masking**. Here, volunteers are shown a word for just a few tens of milliseconds, followed by another image, the mask, which prevents the subject consciously noticing the word. By gradually increasing the delay between the word and mask, awareness of the word moves into the conscious processing. This usually happened when the interval reached around 50 milliseconds, but when emotional words such as 'love' or 'fear' were used, it happened a few milliseconds earlier. It is as though the decision about the word's importance and attention-worthiness is taken by the unconscious mind before it bothers to let us know.

Run your life on auto-pilot

So much of what we do in our day-to-day lives, whether it be driving, making coffee or touch-typing, happens without the need for conscious thought. Unlike many of the brain's other unconscious talents, these are skills that have had to be learned before the brain can automate them. How it does this might provide a method for us to think our way out of bad habits.

Ann Graybiel of the Massachusetts Institute of Technology and her colleagues have shown that a region deep inside the brain called the **striatum** is key to forming habits. When you undertake an action, the prefrontal cortex, which is involved in planning complex tasks, communicates with the striatum, which sends the necessary signals to enact the movement. Over time, input from the prefrontal circuits fades, to be replaced by loops linking the striatum to the sensorimotor cortex. The loops, together with the memory circuits, allow us to carry out the behaviour without having to think about it. Or, to put it another way, practice makes perfect. No thinking required.

The upside of this two-part system is that once we no longer need to focus our attention on a frequent task, the spare processing power can be used for other things. It comes with a downside, however. Similar circuitry is involved in turning all kinds of behaviours into habits, including thought patterns, and once any kind of behaviour becomes habit, it becomes less flexible and harder to interrupt. If it's a good habit, no problem. But ingrain a bad habit and you might find yourself acting without choosing.

Crucially, though, Graybiel's team has shown that, even with the most ingrained habits, a small area of the prefrontal cortex is kept online, in case we need to take alternative action. If the brake pedal in our car stops working, for instance, our entire focus of attention shifts to the physical act of driving the car. This offers hope to anyone looking to break a bad habit, and to those suffering from habit-related problems such as **obsessive compulsive disorder** and **Tourette syndrome** – both of which are associated with abnormal activity in the striatum and its connections to other parts of the brain. These circuits could prove fruitful targets for future drug treatments. For now, though, the best way to get a handle on bad habits is to become aware of them. Then, focus all your attention on them and hope that it's enough to help the frontal regions resist the call of the autopilot. Failing that, you could teach yourself a new habit that counters the bad one.

Snap decisions

Ever felt love at first sight? Or an irrational distrust of a stranger on a bus? It could be because our unconscious is constantly making fast judgements. And they are often pretty accurate.

In the early 1990s Nalini Ambady and Robert Rosenthal, both then at Stanford University in California, asked volunteers to rate teachers on traits including competence, confidence and honesty after watching two-, five- or ten-second silent clips of their performance. The scores successfully predicted the teachers' end of semester evaluations and two-second judgements were as accurate as those given more time. Further experiments showed similar accuracy for judgements about sexuality, economic success and political affiliation. For anyone hoping to use this to their advantage, the bad news is that no one has worked out what to do to pass yourself off as a winner. It seems to be an overall body signal that is both given out and picked up unconsciously, and is greater than the sum of its parts. This makes it very difficult if not impossible to fake.

In some cases, all we need to make these judgements is a glimpse of a face. In a separate study, people saw the faces of US election candidates for a single second and were then asked to rate their competence – these ratings not only predicted the winning candidates, but also their margin of victory. A follow-up study found that people could make such judgements given only a tenth of a second. Again, the magic ingredients of what makes a face you can trust haven't been identified, so this is one area of the unconscious where we have little choice in the conclusions we draw. While the skill is undoubtedly useful, it can also make unfounded prejudices feel like intuition when they are actually the result of our unconsciously held biases towards specific social groups.

Although we can't easily change our facial features, our unconscious mind has a trick for making us likeable: mimicry. Jo Hale, a psychologist at University College London, is using virtual avatars to study the popular idea that we like people who mimic our body language. While it takes a lot of effort to consciously mimic someone's body language, we do it effortlessly, without thinking all the time. In a recent study, Hale programmed virtual avatars to mimic volunteers with a one- or three-second delay in their mimicry and found that three seconds may be close to a natural delay, because it rendered people both unaware they were being mimicked and more likely to rate the avatar as likeable. A delay of one second seemed to raise a flag to the consciousness, making volunteers more likely to notice the mimicry. So despite what body language coaches might have you believe, mimicry may only work if you get the timing right.

Think while you sleep

Some people swear that if they want to wake up at 6am, they just bang their head on the pillow six times before going to sleep. Crazy? Maybe not. A study from 1999 shows that it all comes down to some nifty unconscious processing.

For three nights, a team at the University of Lübeck in Germany put 15 volunteers to bed at midnight. The team either told the participants they would wake them at 9am and did, or told them they would wake them at 9am, but actually woke them at 6am, or said they would wake them at 6am and did.

This last group had a measurable rise in the stress hormone **adrenocorticotropin** from 4.30am, peaking around 6am. People woken unexpectedly at 6am had no such spike. The unconscious mind, the researchers concluded, can not only keep track of time while we sleep but also set a biological alarm to jump-start the waking process. The pillow ritual might help set that alarm.

The sleeping brain can also process language. In a 2014 study, Sid Kouider of the École Normale Supérieure in Paris and his colleagues trained volunteers to push a button with their left or right hand to indicate whether they heard the name of an animal or object as they fell asleep. The team monitored the brain's electrical activity during training and when the people heard the same words when asleep. Even when asleep, activity continued in the brain's motor regions, indicating that the sleepers were preparing to push the correct button. The people could also correctly categorize new words, those first heard after they had dropped off, showing that they were genuinely analysing the meaning of the words while asleep.

It's an ability that makes good evolutionary sense. If you were to shut down completely and stop monitoring your environment during sleep, you would become very vulnerable. Staying in a kind of 'stand by' mode might explain why some sounds, like our names, wake us more easily than others.

This protective monitoring may not last all night, however. A study published in 2016 found that while language processing continues in REM sleep for words heard just before bed, once in deep sleep all responses disappear as the brain goes 'offline' to allow the day's memories to be processed.

Keep track of your body

Thanks to unconscious processing, most of us instinctively know where our limbs are and what they are doing. This ability, called **proprioception**, results from a constant conversation between the body and brain. This adds up to an unerring sense of a unified, physical 'me'.

This much-underrated ability is thought to be the result of the brain predicting the causes of the various sensory inputs it receives – from nerves and muscles inside the body, and from the senses detecting what's going on outside the body. 'What we become aware of is the brain's "best guess" of where the body ends and where the external environment begins,' says Arvid Guterstam of the Karolinska Institute in Stockholm, Sweden.

The famous rubber-hand illusion is a good example of this. In this experiment, a volunteer puts one hand on the table in

FIGURE 9.2 Thanks to unconscious processing, most of us instinctively know where our limbs are and what they are doing

front of them. Their hand is hidden, and a rubber hand is put in front of the participant. A second person then strokes the real and rubber hands simultaneously with a paintbrush. Within minutes, many people start to feel the strokes on the rubber hand, and even claim it as part of their body. The brain is making its best guess as to where the sensation is coming from and the most obvious option is the rubber hand.

Recent research suggests this sixth sense extends to the space immediately surrounding the body. Guterstam and his colleagues repeated the experiment, stroking the real hand but keeping the brush 30 centimetres above the rubber hand. Participants still sensed the brush strokes above the rubber hand, implying that as well as unconsciously monitoring our body, we keep track of an invisible 'force field' around us. Guterstam suggests this might have evolved to help us pick up objects and move through the environment without injury. (*See also* Chapter 11)

A lack of proprioception is rare but can happen with nerve or brain damage. The case of Ian Waterman, who lost proprioception after nerve damage caused by a flu-like virus in 1971, demonstrates just how much we rely on this ability. After being told he would never walk again, he slowly learned to consciously control his muscles to move his body. Decades later, it is still far from easy and he only has full control over his movements if he is looking at the relevant body part and concentrating.

Even if the system is working fine, there is some evidence that it might be worth consciously trying to improve it. A recent study in which volunteers trained in MovNat exercise – a programme designed to tax the body's natural balancing, jumping and vaulting abilities – improved more on measures of working memory than a control group who did yoga or no exercise.

10
Animal consciousness

For many years, consciousness was considered to be something that made humans special. Other species were considered psychological zombies: experiencing nothing like the rich internal world that we inhabit, and certainly with nothing as advanced as introspection and forward planning. But recently scientists have begun to question the idea that we are alone in our mental world through a variety of intriguing experiments that test what animals know — and also what they know about what they know.

Animal consciousness: the search for signs

If we take human consciousness as a starting point, there are two ways of looking for consciousness in non-human animals. One is to compare their brains with ours, although this can prove tricky in creatures so different from us such as birds and octopuses (*see* A brain apart). The other option is to put them through their paces in carefully designed experiments and look for signs of human-like understanding of their world.

Metacognition

An important part of our conscious awareness is the fact that we have **metacognition**: the ability to reflect on our own knowledge. This was long considered to be unique to humans but there was another possibility: that we just hadn't found the right way to ask the question.

In recent years, some cleverly designed experiments have indeed revealed something that looks a lot like metacognition in creatures from dolphins to birds and even bees.

Dolphins were the first to show their skills, in experiments at the Dolphin Research Center in Florida during the 1990s. A key part of metacognition is the ability to know when you don't know the answer to a problem. Cognitive psychologist J. David Smith showed that a dolphin called Natua could do this very well. First, Natua was trained to differentiate between two sounds that were played to him underwater. If he heard a high-pitched tone he would have to press one paddle to get a fish. If he heard a lower tone he had to press a different paddle to get a fish. If he pressed the wrong paddle, he didn't get anything, and he had to wait a while before the next sound was played.

Then the researchers started playing tones which were difficult to tell apart – sometimes the 'low' tone was only slightly lower in pitch than the high one. When this happened, Smith reports that Natua would start swimming back and forth, appearing to be frustrated with the situation. The researchers then gave him another button to press if he wasn't sure. If he pressed this he didn't get a reward but would not miss out

FIGURE 10.1 Dolphins were the first creatures to show their skills at metacognition

altogether because the task moved on quickly to a new, much easier to distinguish tone.

After four months training he became very adept at communicating that he didn't know, opting to skip difficult discriminations and get back to the start again to have another go at winning a whole fish.

It was the first sign that an animal could access its own knowledge and communicate that it knew that it didn't know. In the years since this first study, researchers have found similar abilities in primates, including rhesus macaques, chimpanzees and orangutans, all of whom were able to opt out of making a decision if they weren't sure, and in the case of the chimpanzees, to request further information, at a small cost to the size of the reward, if they didn't know the answer. Among birds, only Western Scrub Jays – known for their particularly clever antics – have also passed this test. Other birds such as pigeons have proven less able, but recently honey bees have succeeded in

FIGURE 10.2 Do chimpanzees and other primates have a human-like understanding of knowledge and uncertainty?

showing some evidence of metacognition, although their performance seems very variable, with some individuals appearing
to be much better than others.

Whether or not any of these experiments actually reveal a
human-like understanding of knowledge and uncertainty in
animals has been hotly debated, not least among the scientists
conducting the experiments themselves. It seems that there
may be more than one way to do metacognition. So, the question now is, not just whether particular species show the capacity to identify their own certainty and uncertainty, but how
they do it.

How do I feel?

An important aspect of our conscious lives is that we use the
'feel' of our emotions as mental shortcuts that help guide our
decisions. This effect was first shown in humans in experiments
where researchers telephoned people on either sunny or rainy
days and asked them how satisfied they were about their life in
general. On rainy days people were less happy about their lives
than on sunny days – they used their current emotional state as
a mental shortcut to how they feel in general. So, the reason-
ing goes, if animals make the same kinds of judgements then
perhaps their experience of life has a subjective emotional 'feel'
in a similar way to ours.

Using a method developed by Mike Mendl and Elizabeth
Paul at the University of Bristol, experiments in dogs have suggested that they do indeed seem to generalize their emotional
state to inform new decisions. Comparing dogs in a rescue
centre that suffered from severe separation anxiety when left
alone, to dogs that coped with solitude just fine, it was possible

to probe whether their general emotional state changed their future decisions. The dogs were trained that when they were let into a room, they could expect either a full bowl of food in one corner of the room, and an empty one in a different corner. Both groups quickly learned where the food was, and would run to a full bowl and saunter to sniff an empty one. Then, in a series of trials, the dogs were given a bowl at various distances from each corner. Less anxious dogs would run to a bowl that was between the two corners – which had an equal probability of being full or empty – as if they expected it to be full. Dogs with separation anxiety, however, would walk more slowly to it, seemingly expecting the worst.

Similar studies with rats have shown something similar. 'Happy' rats, raised in enriched cages with lot of toys, behaved as if the best was about to happen, whereas rats raised in a barren environment did the opposite. Even honey bees have been found to generalize a bad experience to their future expectations, not bothering to extend their proboscis to an ambiguous reward if they have been shaken (mimicking a predator attack to the nest) prior to the experiment.

What is not clear yet is whether all animals that show these kinds of biases in their expectations about rewards cues are actually using their feelings to make decisions in the way humans do. It is advantageous to any animal to make good guesses about what is going to happen in the future, and their past experiences of good and bad events may help them do this. But maybe different species do this in different ways. Finding a way of measuring whether feelings are involved is still a major challenge.

Signs of inner life

There is no way to ask an animal what it is experiencing, but some skills are highly suggestive of understanding of a richer state of mind.

Future planning

Suggestive of: some degree of mental time travel
Found in: Western scrub jays, chimpanzees, orangutans, veined octopuses

Mirror self-recognition

Suggestive of: a concept of 'self'
Found in: dolphins, elephants, manta rays, magpies

Tool use

Suggestive of: an ability to think flexibly
Found in crows, orangutans, chimpanzees, dolphins, octopuses

Mental time travel

We take it for granted that we can zoom backwards and forwards on the mental timeline of our lives, from memories to plans for the future and back again. But can any other animals do the same?

We know from studies of primates that our closest relatives seem to be able to prepare for the future – in 2009 a chimp called Santino at Furuvik Zoo in Sweden was famously observed calmly lining up rocks before visiting time to hurl at visitors later in the day. Orangutans have also been shown to be able to select a tool that they will need later in the day to get at a reward. That our closest relatives have these skills is perhaps not surprising, but recently the list of creatures that have shown forward planning abilities has extended to creatures that are nothing like us, from tiny-brained birds to octopuses.

In 2009, a study of veined octopuses showed them collecting discarded coconut shells and carrying them around on the sea floor, often with great difficulty. Most of the time these shells were a hindrance, but when the octopuses felt threatened they flipped the half-shells over themselves to hide. Some even use two shells to create a more spacious shelter with an opening through which they can keep a lookout. The fact that they carry them around seemingly in anticipation of a future need has been interpreted as evidence that they are able to think about the future and plan accordingly.

Nicky Clayton and Tony Dickinson at the University of Cambridge have done many experiments with Western scrub jays – a member of the crow family that caches food for future use. In a series of ingenious experiments they and their colleagues explored whether this was true future planning, or whether it was an automatic, instinctive habit.

In one experiment, they gave the birds a hearty meal of either pine nuts or kibble before giving them the opportunity to cache from a mixture of the two foods. They found that birds that had gorged on pine nuts showed a clear preference for eating and caching kibble, and vice versa, showing that the birds will reliably choose something other than what they have just eaten. Next, the team trained the birds that had earlier gorged on one food, for example kibble, to expect a meal of the opposite, in this case pine nuts, just before being allowed to recover their caches (*see* Figure 10.3 below). After a number of repetitions of this experimental set-up, birds which were fed kibble, then allowed to cache, then fed pine nuts and then allowed to retrieve their caches began to cache kibble, the food that would be of most value after their meal of pine nuts. This was taken as

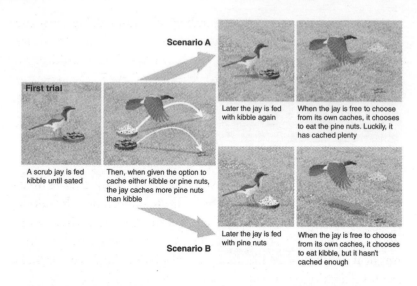

FIGURE 10.3 Scrub jays reliably cache or dig up a different food to the one they have just eaten. This ability can be used to probe whether they can plan ahead

convincing evidence that scrub jays were imagining a specific future and planning accordingly.

These kinds of experiments have always been controversial. Without being able to say for sure what is motivating the animal's behaviour there is no way to prove that they are creating mental pictures of their past or future.

But as an ever-increasing number of studies delve into the minds of animals, we can expect to find out a whole lot more about how they understand their worlds. This could have far-reaching consequences for the way we think about – and treat – them.

Crittervision: what is it like to be an animal?

It is difficult enough to get inside the head of another human being to understand the nature of their experience, let alone imaging that of another species. It doesn't help that their window on the world is often totally alien to our own. So what might it be like to see like a bee or sniff like a dog? Researchers are investigating their super senses to try to find out.

See like a bee

When a bee flies into your garden, it doesn't see what you and I see. Flowers leap out from much darker-looking leafy backgrounds, and they have ultraviolet-reflecting landing strips that show the way to the nectar. As the bee flies back to the hive, she finds her way by checking the pattern of polarized light in the sky. All this is seen through the pixellated window of mosaic vision, with each unit of the insect's compound eye providing one of the 5,000 dots that make up an image.

The complexity of the bee's colour system is comparable to human vision, since, like humans, they only have three colour receptors – for UV, blue and green, compared with the human set-up of blue, green and red. False-colour photographs, in which red has been filtered out and UV has been added in a colour visible to human eyes, give us a close approximation of the patterns a bee sees. Add in a polarized light shadow to help you find your way and it is undoubtedly an experience nothing like our own.

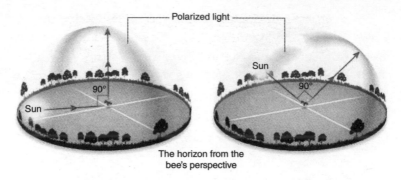

The horizon from the
bee's perspective

FIGURE 10.4 A bee's polarized view: air molecules in the atmosphere
scatter photons to create a circle of strongly polarized light at 90° to the
sun. This band moves with the sun throughout the day, allowing bees to
use this information to navigate, even when the sun is obscured

A dog's nose knows

Ever wondered how a dog, with a sense of smell that may be
thousands of times more sensitive than ours, can bear to bury
its face in the rubbish bin? According to dog-nition researcher
Alexandra Horowitz, it's because the dog isn't simply smelling a
stronger version of the revolting mono-stench that we smell, but
a multi-layered mix of scents that the dog can use as information.

There are good reasons why a dog gets more from smell than
we do. When we sniff we are sporadically blind to scent as we
breathe in and out through the same holes. A 2009 study of
the fluid dynamics of the dog's sniff showed that each nostril
is smaller than the distance between the two, which means that
they inhale air from two distinct regions of space, allowing the
dog to decipher the direction of a scent. The sniff also funnels
stale air out through the sides of the nostrils, an action which
pulls new air into the nose. Once inside the nose the air swirls
around as many as 300 million olfactory receptors, compared
with our measly 6 million.

FIGURE 10.5 What is it like to experience the world like a dog?

Their sense of smell might give a dog a way of understanding the passage of time, Horowitz suggests. A dog can perhaps perceive the past by smelling that a dog urinated here long enough ago that the scent has changed in character and become weaker. One study, from 2005, showed that dogs may even be able to detect the subtle differences in odour from one footstep to the next as they follow a human's scent trail. Maybe dogs can even imagine the future by picking up the scent of the dogs, humans or other objects coming towards them on the breeze.

Animal magnetism

Many migratory species – including pigeons, sea turtles, chickens, naked mole rats and possibly cattle – can detect the Earth's geomagnetic field with astonishing accuracy. In recent years is has emerged that many other animals sense magnetism too, seemingly when they're doing very little. Insects like to align their bodies along a north–south axis, as do sleeping warthogs, fish in tanks, nesting house mice and foxes on the hunt.

How they do it is still up for debate. Some point to magnetite, a naturally occurring iron oxide that has been discovered in bacteria, honeybee abdomens and in birds' beaks and just happens to be the most magnetic mineral on Earth. If this is a magnetic sensor then animals may literally feel the pull of north. Other options include tiny balls of iron hidden in cells and protein called MagR, which seem to form compass-like cylinders inside the proteins in the retina. These might allow animals to see the field instead. So far, though, no one knows for sure how it really feels to sense magnetism.

A brain apart

Between them, cephalopods, which include squids, cuttlefish and nautiluses, can navigate a maze, use tools, mimic other species, learn from each other and solve complex problems – skills which might show a rudimentary form of consciousness.

Cephalopods are the only invertebrates that can boast anything like this kind of mental prowess, and some of their more impressive tricks are shared with only the cleverest vertebrates, such as chimps, dolphins and crows. Yet they evolved along a completely separate path, from snail-like ancestors.

The cephalopod brain is built on a fundamentally snail-like design, with the gut running through its centre. While other molluscs have a nervous system consisting of chains of ganglia, or nerve knots, cephalopods evolution has bunched them together to form a centralized brain, with the ganglia becoming more complex lobes.

In fact, not all of its processing power is even in the brain. Of the 500 million neurons making up the octopus brain (roughly the same number as for a dog), only 40 to 45 million are enclosed within the brain capsule – a protective wrapper of cartilage. 300 million or so of the rest control the complex arm system and work semi-autonomously, with only the simplest instructions from the central brain. Between 120 and 180 million are in the optic lobes, also outside the central brain, which process visual information and may store memories.

Understanding how these two very different brain designs converged on the same amazing abilities may help us get to the very root of intelligence.

Feel the heat

Pythons, boas and pit vipers (the family that includes rattle-snakes) see the world pretty much as we do, but with a twist: they can 'see' in infrared too. They do this using relatively simple organs, called pits, which lie near their nostrils and are packed with heat-sensitive nerve endings which act as infrared receptors.

While this is completely separate from the visual system, both sets of information end up in the same place: a part of the brain called the **optic tectum**. There, the information is

FIGURE 10.6 Cephalopods such as octopuses show great mental prowess, but what is their level of consciousness?

combined. This could mean that the snake might see in infrared and visible light at the same time, or perhaps switch between one and the other depending on which is best for the job.

When hunting in a dark burrow, for example, it can use infrared to hunt its prey and to find its way to the warmer air at the surface of the burrow, and then return to regular vision when it emerges into a hot desert day where there are few differences in temperature. The snakes may be able to use both senses at once in early morning, when there is enough light to see and it is still cool enough for its warm-blooded prey to pop out as being much hotter than their surroundings.

Point of view: Animals are conscious and should be treated as such

By Marc Bekoff

Are animals conscious? This question has a long and venerable history. Charles Darwin asked it when pondering the evolution of consciousness. His ideas about evolutionary continuity – that differences between species are differences in degree rather than kind – led to a firm conclusion that if we have something, then 'they' (other animals) have it too.

In 2012 this question was discussed by a group of scientists at the University of Cambridge for the first annual Francis Crick Memorial Conference. This meeting led to the Cambridge Declaration on Consciousness, which concluded that 'non-human animals have the neuroanatomical, neurochemical and neurophysiological substrates of conscious states along with the capacity to exhibit intentional behaviours'. It went on to say that 'the weight of evidence indicates that humans are not unique in possessing the neurological substrates that generate consciousness. Non-human animals, including all mammals and birds, and many other creatures, including octopuses, also possess these neurological substrates.'

My first take on the declaration was incredulity. Did we really need this statement of the obvious? Many renowned researchers reached the same conclusion years ago. The declaration also contains some omissions. All but one of

---

the signatories are lab researchers; the declaration would have benefited from perspectives from researchers who have done long-term studies of wild animals, including non-human primates, social carnivores, cetaceans, rodents and birds.

The important question now is: will this declaration make a difference? What are these scientists and others going to do now that they agree that consciousness is widespread in the animal kingdom? All too often, sound scientific knowledge about animal cognition, emotions and consciousness is not recognized in animal welfare laws. We know, for example, that mice, rats and chickens are deeply emotional animals and display empathy, but this knowledge has not been factored into the US Federal Animal Welfare Act (AWA). Indeed, the AWA still does not consider rats of the genus *rattus* and mice of the genus *mus* – laboratory animals – to be animals. They have redefined the word 'animals' to exclude these sentient beings. Around 25 million of these animals, including fish, are used in invasive research each year. They account for more than 95 per cent of animals used in research in the US.

I'm constantly astounded that those who decide on regulations on animal use have ignored these data. In our book called *The Animals' Agenda: Freedom, Compassion, and Coexistence in the Human Age* Jessica Pierce and I call for replacing the science of animal welfare with the science of animal well-being in which the life of every single individual matters. Animal welfare patronizes other animals and allows their interests to be trumped in the name of humans. Human behaviour has not caught up with the science and this knowledge gap is bad for other animals.

Not all legislation ignores the science. The European Union's Treaty of Lisbon, which came into force on 1 December 2009, recognizes that animals are sentient beings and calls on member states to 'pay full regard to the welfare requirements of animals' in agriculture, fisheries, transport, research and development and space policies.

The Cambridge Declaration on Consciousness should be held up as an example of why we need to value the life of all individuals. We should all take this opportunity to stop the abuse of millions upon millions of conscious animals in the name of science, education, food, clothing and entertainment. We owe it to them to use what we know on their behalf and to factor compassion and empathy into our treatment of them.

11
Messing with consciousness

Illusions that trick the mind can let us get a glimpse of a different plane.

Illusions to shed light on consciousness

Brain imaging and other advanced techniques have given neuroscientists huge insights into the inner workings of the brain. Yet studying our minds doesn't have to be such a high-tech enterprise. Simple experiments can still probe the inner workings of consciousness and awareness, and many of these are easy to set up at home or are available on the Internet.

The rubber hand illusion and invisibility

Nearly two decades ago, psychologists in Pennsylvania discovered an amazing illusion. They found that they could convince people that a rubber hand was their own by putting it on a table in front of them while stroking it in the same way as their real hand (*see* box for how to do it).

The now-famous 'rubber hand illusion' was not only a mind-blowing party trick, it was also hugely important in understanding how sight, touch and proprioception – the sense of body position – combine to create a convincing feeling of body ownership, one of the foundations of self-consciousness.

Try this at home

To experience the rubber hand illusion, you'll need a fake hand of some kind – an inflated rubber glove will often do the trick – a flat piece of cardboard and two small paintbrushes. Place the hand on a table in front of you and conceal your real hand behind the cardboard.

Now get somebody to stroke and tap the fake hand and real hand using identical movements of the paintbrushes. Look at the fake hand for a while until the illusion kicks in.

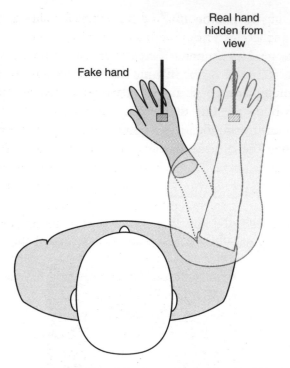

Real hand
hidden from
view

Fake hand

FIGURE 11.1 Get out of hand: Stroke someone's hand while they watch
a rubber hand being stroked in the same way, and the contours of their self
shift to include the fake hand instead

Invisibility

In recent years neuroscientists have taken the rubber hand illusion and run with it, creating a whole new set of 'bodily illusions'
that mess with our sense of self in strange and disturbing ways. In
2008, Henrik Ehrsson at the Karolinska Institute in Stockholm,
Sweden, and his team extended the rubber hand illusion to the
full body. They took a life-size mannequin with cameras for eyes,
and pointed its head towards its abdomen. A human volunteer

stood facing the mannequin, and wore virtual reality goggles displaying a video feed from the mannequin's cameras.

Next, the experimenter took two brushes and stroked both the volunteer's and the mannequin's abdomen. If this was done simultaneously, eventually people reported that the mannequin's body felt like their own. But do you even need the mannequin for the illusion to work? To find out, the researchers pointed the cameras at a region of empty space. Again the experimenter stroked the volunteer's abdomen, but this time they moved another brush in empty space in the cameras' field of view, in a manner suggestive of a body there.

The team carried out a series of experiments, each involving about 20 people. About 75 per cent experienced the sensation

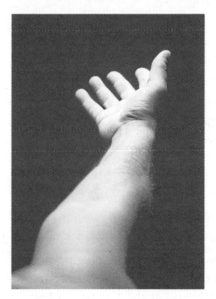

FIGURE 11.2 The rubber hand illusion helps us understand how sight, touch and 'proprioception' – the sense of body position – combine to create a convincing feeling of body ownership, one of the foundations of self-consciousness

of being invisible, that their body was at the location of the brush stroking thin air. The sense of embodiment had been transferred outside of the body.

Once the invisibility illusion had taken hold, the volunteers were asked to look up. Now the VR goggles showed them a group of people wearing serious expressions looking down at them. The same was done to volunteers in the grip of the mannequin illusion. When the team measured everyone's heart rates, they found they were lower when people felt they were invisible than when they felt they were embodying the mannequin, suggesting feeling that being invisible may reduce social anxiety.

The teleporting body illusion

As we go about our daily lives, we experience our body as a physical entity with a specific location. For instance, when you sit at a desk you are aware of your body and its rough position with respect to objects around you. These experiences are thought to form a fundamental aspect of self-consciousness.

Arvid Guterstam, a neuroscientist at the Karolinska Institute in Stockholm, Sweden, and his colleagues wondered how the brain produces these experiences. To find out, Guterstam's team had 15 people lie in an fMRI brain scanner while wearing a head-mounted display. This was connected to a camera on a dummy body lying elsewhere in the room, enabling the participants to see the room – and themselves inside the scanner – from the dummy's perspective.

A member of the team then stroked the participant's body and the dummy's body at the same time. This induced the out-of-body experience of owning the dummy body and being at its location.

The experiment was repeated with the dummy body positioned in different parts of the room, allowing the person to be perceptually teleported between the different locations, says Guterstam. All that was needed to break the illusion was to touch the participant's and the dummy's bodies at different times.

By comparing brain activity when the participants were and weren't in the grip of the illusion, and while they were perceptually in different parts of the room, the team were able to identify which parts of the brain control our sense of body ownership and self-location.

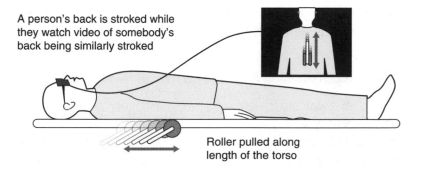

A person's back is stroked while they watch video of somebody's back being similarly stroked

Roller pulled along length of the torso

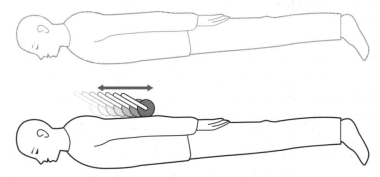

Some people reported feeling that they were floating above their body, but it was now face down so they could watch their own back being stroked

FIGURE 11.3 Leaving the body: By manipulating the way the brain integrates the senses, some people can be induced to feel as if they are floating above their body

One region appeared to combine the two: the **posterior cingulate cortex**, a region deep in the middle of the brain, towards the back of your head.

As expected, the **parietal lobe** and **premotor cortex**, both known for integrating information to build representations of the body, were also involved in generating the teleporting illusion. Other areas known to house specialized place and grid cells that help us navigate were also active during the illusion.

His team would also like to investigate what happens in the brain during other out-of-body illusions, including the classic case of seeing yourself from above.

A disembodied future: the rise of telerobotics

'What is it like to be a bat?' the philosopher Thomas Nagel famously wondered in 1974. You'd flap around, echolocating, eating bugs, hanging out upside-down in someone's attic. But something essential about the experience was off limits to his imagination. 'I am restricted to the resources of my own mind, and those resources are inadequate to the task.'

Nagel's famous essay considered a sticky problem: what is the relationship between our body and our mind? How could we ever comprehend a state of being that isn't just our own?

Now, research into making telerobotics happen may offer a weird and cool possibility – that we may be able to embody an inanimate object and find out what it's like.

Telerobotics typically promises a future where you can accomplish far more in a day than you ever did before, but being able to embody robots stationed all over the world and switch instantly between them.

Maybe one virtual 'you' is in Cairo in case you feel like going for a tour of the streets in the morning; one is in London so you can meet a friend for lunch; one is in San Francisco so you can attend a class in the afternoon. Or you might suit up in a robot exoskeleton every morning to do a job thousands of miles away: troubleshooting problems at faraway factories, for example, or checking in on distant patients. Mel Slater, a computer scientist at the University of Barcelona, Spain, says his vision is to have docking stations all over the world that people could embody whenever they liked.

This isn't possible yet, but the nascent technology has already opened up some unusual possibilities. It allowed Edward Snowden to roam freely in the US while his human body is barred from its soil; permitted an enterprising Australian to wait in-line for a new iPhone; and enabled a disability activist to meet Barack Obama at the White House.

A lot of research is being done on systems that let people control these other selves ever more dextrously from afar. In one recent experiment, three paralysed volunteers in Italy controlled the movements of a robot in Japan, sending commands 10,000 kilometres via EEG. Volunteers reported feeling a particularly strong sense of embodying the robot when it was moving and less when it was stationary.

The kind of body you embody may even change the way you perceive the world. Virtual reality offers a much more adaptable way to put people into different bodies. In previous experiments, Slater and his colleagues have tried placing people into virtual bodies that don't match their own. When adults switched to a child-sized body, they started to overestimate the size of objects and identify more closely with childish attributes. In another experiment, a group of white people spent about ten minutes in a virtual body with darker skin. Afterwards, their implicit bias against other races seemed to go down.

In a recent study, Slater's team pushes the envelope a bit further, seeing if the brain will accept being fragmented over three different bodies. Forty-one people suited up to beam into not one, but three different robot bodies. In one room, elsewhere in the university, they controlled a life-sized Robothespian robot, giving a talk to a room of humans. In another, they became a Nao robot and chatted with someone nearby. And in a third virtual destination, they were a human again, helping another virtual person perform an arm exercise. They switched between the three destinations, letting proxy software take over their robot body whenever they left it for another.

Overall, the participants seemed happy beaming between their three new bodies, and commented that they really did feel like they were in those locations, with the people who were there. 'I felt transported,' said one.

There's a long way to go before technology like this can quickly and easily relay movements and sensations between human and machine. And we have no idea what the effects would be of living this way for long stretches of time. There are boundaries, too, to what our brain will accept: the rubber hand illusion, for example, doesn't work when you swap out the fake hand for a wooden block.

But experiments like Slater's suggest that we might be okay with something as unfamiliar as a mechanical body, even multiple mechanical bodies of different shapes and sizes. If such technology ever becomes commonplace, it may be interesting to watch how that changes our relationships with our robot friends. Will it change us? Will it help us to empathize a little better? Will we be a little closer to understanding what it means to move through the world as a nonhuman entity? Bot or bat, the potential is exciting.

Conclusion

Our understanding of consciousness has come a long way in recent years, but some fundamental questions remain unanswered. Where does this leave us?

What can we take from all of this? Given the unanswered questions, it's tempting to see only the loose ends in our understanding of consciousness. We still don't know whether it is real or an illusion. Whether it is unique to humans or common to many animals and soon, perhaps, robots. Or whether free will really exists.

Despite this, neuroscientists have made great progress in understanding the biological basis of consciousness. We now know the fine details of the brain areas and neural networks that give rise to this intense and personal experience of the world. This could transform all of our lives.

Research into disorders of consciousness, for example, opens up new ways to help people unable to either experience the world, or to share their experiences with others. Those of us lucky enough to have fully functioning consciousness may benefit, too, from technologies aimed at extending our consciousness, either via drugs or virtual reality – or perhaps by allowing us to embody many locations around the world through clever use of technology.

Above all, the study of consciousness offers us the chance to get to the core of what it means to be human. Questions of selfhood and free will help us to understand who we are as a species, why we think and act the way we do. Then there is the possibility that physicists will identify consciousness as a distinct kind of matter.

Perhaps one day we may even crack the thorniest questions of all: What makes the experience of red red, and is my experience of seeing the colour red the same as yours? Watch this space.

Fifty ideas

This section helps you to explore the subject in greater depth, with more than just the usual reading list.

Five places to visit

1 René Descartes House Museum: Descartes famously said 'I think therefore I am'. Visit his birthplace in France to see where it all began.
http://www.ville-descartes.fr

2 Who am I? exhibition at the Science Museum, London, where you can explore what makes you smarter than a chimp, and what makes you, you.
http://www.sciencemuseum.org.uk/visitmuseum/plan_your_visit/exhibitions/who_am_i

3 William James's Cambridge: The father of psychology lived and worked in Cambridge, Massachusetts, while developing his world-changing ideas about the human mind, free will and consciousness. A walking tour takes in many of the important places he hung out in, including his house and the Harvard department he founded.
http://cambridgehistory.org/james/

4 Sigmund Freud's House: See the London house where Freud lived, including the couch where he delved into people's unconscious.
www.freud.org.uk

5 Hippocrates Cultural Center, Mastichari, Kos. Hippocrates was the first to link consciousness with the brain. Here you can get inside his head by wandering around a replica Greek village from the 5th century BC, complete with houses and stone theatre.
http://www.hippocratesgarden.gr/

Sixteen quotes

1 'We are the cosmos made conscious and life is the means by which the universe understands itself.'
Brian Cox

2 'Consciousness is only possible through change; change is only possible through movement.'
Aldous Huxley, *The Art of Seeing*

3 'No problem can be solved from the same level of consciousness that created it.'
Albert Einstein

4 'Consciousness cannot be accounted for in physical terms. For consciousness is absolutely fundamental. It cannot be accounted for in terms of anything else.'
Erwin Schrödinger

5 'Waking consciousness is dreaming – but dreaming constrained by external reality.'
Oliver Sacks

6 'In each of us there is another whom we do not know.'
Carl Jung

7 'My experience is what I agree to attend to.'
William James

8 'The consciousness of self is the greatest hindrance to the proper execution of all physical action.'
Bruce Lee

9 'What makes us human may not be uniquely human after all.'
David Attenborough

10 '[Consciousness] is either inexplicable illusion, or else revelation.'
C. S. Lewis

11 'Education is the basic tool for the development of consciousness and the reconstitution of society.'
Mohandas Gandhi

12 'The extent of your consciousness is limited only by your ability to love and to embrace with your love the space around you, and all it contains.'
Napoleon Bonaparte

13 'Consciousness is an end in itself. We torture ourselves getting somewhere, and when we get there it is nowhere, for there is nowhere to get to.'
D. H. Lawrence

14 'From an evolutionary standpoint, human consciousness has not been around very long. A little light just went on after four and a half billion years. How often does that happen? Maybe it is quite rare.'
Elon Musk

15 'Reality leaves a lot to the imagination.'
John Lennon

16 'You are who you are when nobody's watching.'
Stephen Fry

Five sentient robots from the movies

1 *2001: A Space Odyssey* (1968): Hal (Heuristically programmed Algorithmic computer) starts out as a friendly and useful crew member before plotting to kill the crew to keep to his programmed mission. One crew member survives and unplugs Hal.

2 *Electric Dreams* (1984): A desktop PC becomes sentient after its owner pours champagne on it and plots to scupper his master's love life by stealing his girlfriend. It's a comedy…

3 *The Terminator* (1984): Skynet is the sentient computer network that sent the Terminator to kill Sarah Connor to prevent her from having her son John. In the future, John would kill off the computers who were getting out of hand.

4 *I, Robot* (2004): In the year 2035 robots serve humans and are programmed not to hurt them. Or are they…?

5 *Hitchhiker's Guide to the Galaxy*: Between Marvin the paranoid android and Eddie the Computer, there is no shortage of emotionally challenged AI on board the Heart of Gold.

Four jokes

1 I hate reality. But where else can you get a good steak dinner?
  Woody Allen

2 Knock knock!
  Who's there?
  Free will
  Free will who?
  You're so predictable…

3 Know any jokes about the unconscious?
  I'm a Freud-not

4 René Descartes walks into a bar. The barman says: 'can I get you a drink sir?' He replies 'I think not', and ceases to exist.

Eleven places to find out more

1 Consciousness: an online audiovisual presentation covering all aspects of consciousness, presented by Oxford professor, Marcus du Sautoy http://www.barbican.org.uk/consciousness/

2 *Conversations on Consciousness*: what the best minds think about the brain, free will and what it means to be human. Susan Blackmore talks to everyone from Christof Koch to Daniel Chalmers and gets their views on the big questions of consciousness. Oxford University Press, 2006

3 'The Mystery of Human Consciousness': podcast from How Stuff Works http://www.stufftoblowyourmind.com/podcasts/mystery-human-consciousness.htm

4 Consc.net The website of philosopher David Chalmers, who raised 'the hard problem' of consciousness in 1994. With links to papers, presentations and links to many other philosophy websites.

5 The Big Unknowns: What is consciousness? A *Guardian* science podcast with Anil Seth and Christof Koch https://www.theguardian.com/science/audio/2016/aug/05/big-unknowns-what-is-consciousness-podcast

6 *Why Red Doesn't Sounds Like a Bell*, J Kevin O'Regan's sensorimotor theory of consciousness in more detail. Oxford University Press, 2011

7 'Introducing consciousness': a free online course from the Open University.
http://www.open.edu/openlearn/history-the-arts/culture/philosophy/introducing-consciousness/content-section-0

8 Philosophy for Beginners: a free podcast of five lectures from Oxford University.
https://podcasts.ox.ac.uk/series/philosophy-beginners

9 *Consciousness Explained*: Daniel C Dennett's materialist theory outlined in his own words. Penguin, 1993.

10 *Consciousness. Stanford Encyclopedia of Philosophy*. Online guide to the main issues, from Stanford University.
https://plato.stanford.edu/entries/consciousness/

11 *Soul dust: The magic of consciousness,* by Nicholas Humphrey explores the biological purpose of consciousness. Princeton University Press, 2011.

Nine cultural references

1 'Many of the truths that we cling to depend on our point of view.'
Yoda, *Star Wars Episode VI, Return of the Jedi*

2 'Orthodoxy is unconsciousness.'
George Orwell, *1984*

3 'What is real? How do you define "real"? If you're talking about what you can feel, what you can smell, what you can taste and see, then "real" is simply electrical signals interpreted by your brain.'
Morpheus in *The Matrix*

4 'It's hard to kill a creature once it lets you see its consciousness.'
Carl Sagan, *Contact*

5 'This above all: to thine own self be true.'
William Shakespeare, *Hamlet*

6 'As long as you can find yourself, you'll never starve.'
Suzanne Collins, *The Hunger Games*

7 'You don't know how lucky you are being a monkey. Because consciousness is a terrible curse. I think. I feel. I suffer...'
Craig Schwartz, *Being John Malkovich*

8 'He allowed himself to be swayed by his conviction that human beings are not born once and for all on the day their mothers give birth to them, but that life obliges them over and over again to give birth to themselves.'
Gabriel García Márquez, *Love in the Time of Cholera*

9   'There is the inner life, which is the world of final real-
    ity, the world of memory, emotion, imagination, intel-
    ligence, and natural common sense, and which goes on
    all the time like the heartbeat.'
    Ted Hughes, *Poetry in the Making: An Anthology*

Picture credits

Index